W9-CUJ-217

# The DYNAMICS of

# TECHNICAL

# CONTROVERSY

by Allan Mazur

# The DYNAMICS of TECHNICAL CONTROVERSY

by Allan Mazur

Communications Press, Inc.

Grateful acknowledgment is made for permission to quote extensively
from the following:
Mazur, Allan, "Disputes Between Experts," *Minerva,* XI, 2, April 1973,
pp. 243-62, © 1973 by *Minerva.*
Mazur, Allan, "Opposition to Technological Innovation," *Minerva,*
XIII, 1, Spring 1975, pp. 58-81, © 1975 by *Minerva.*
Mazur, Allan, with Marino, Andrew, and Becker, Robert, "Separating
Factual Disputes from Value Disputes in Controversies over
Technology," *Technology in Society,* Vol. 1, 1979, pp. 229-37, © 1979 by
Pergamon Press, Ltd.
Mazur, Allan, and Leahy, Peter, "The Rise and Fall of Public
Opposition in Specific Social Movements," *Social Studies of Science,*
Vol. 10, 1980, pp. 259-84, © 1980 by SAGE Publications Ltd.

Printed in the United States of America

Published by Communications Press, Inc.

Library of Congress Cataloging in Publication Data:

Mazur, Allan.
    The dynamics of technical controversy.


    Includes bibliographical references and index.
    1. Technology — Social aspects.   I. Title.
T14.5.M39        306'.4        81-3257
ISBN 0-89461-033-3        AACR2
ISBN 0-89461-034-1 (pbk.)

To Joseph and David Mazur

# Contents

# List of Exhibits

# Preface

The products of science and technology have increasingly become objects of controversy, and political debates about them have grown so complex that few people understand what is going on. My goal here is to explain these controversies and to suggest how they may be used as effective means for technology assessment.

Most research in this area consists of either case studies or opinion surveys about particular technologies that have been under protest. Nuclear power has been the leading topic in recent years, as was fluoridation in earlier decades. Some of this work is of undoubted high quality (and some is awful), but even the best seems to have reached its limit. The time has certainly arrived for a broader, coordinated, comparative study of technical controversy, lifted above the specifics of any one case so as to provide a general theoretical understanding of the phenomenon and to guide further research with specific hypotheses to be tested. Therefore I am not concerned here with the details of any single controversy, nor in judging which side is right or wrong. Instead I want to suggest regularities in behavior which often occur across the whole class of technical controversies, and to indicate the kinds of data, now lacking, which are required if we are to advance our understanding.

This book has three major shortcomings which I will admit here before the reader discovers them for himself. First, the whole discussion is built on a weak data base, for the qualitative descriptions found in case studies do not lend themselves to tight comparisons across controversies, and superficial questionnaire surveys provide little insight into the dynamics of controversy. Second, given my aim of forming general principles, and the lack of an adequate data base to constrain my speculations, I have probably simplified reality inordinately in order to fit it into neater patterns than actually exist. Third, the discussion is limited almost entirely to affairs in the United States because my resources are too scant for a serious study of technical controversies in the other industrial countries.

I appreciate the critical comments and advice of Joseph Ben-David, Edward Groth, III, Joseph Haberer, Mary Louise Hollowell, Louis

Kriesberg, W. Henry Lambright, Gerald Markle, Robert Mitchell, Dorothy Nelkin, Charles Perrow, Eugene Rosa, Edward Shils, Manfred Stanley, and Albert Teich. I thank Harriet Hanlon, Ruthe Kassel, Nancy Klein, Julie Mazur, Rachel Mazur, Liv Muhre, Linda Richard, and Diane Tardiff for their excellent secretarial assistance.

# 1
# Technical Controversies

When Benjamin Franklin invented the lightning rod, it had been the custom for centuries in Europe to dispel lightning from thunderstorms by ringing church bells. The tenacity of this practice was amazing since high church steeples make excellent targets for lightning. Furthermore, electricity easily travels down a rain-wetted bell rope to the ringer. One eighteenth-century European source reports that over a 33-year period, lightning struck 386 church steeples killing 103 bell ringers.

Some clerics immediately accepted Franklin's rods as an improvement over bell ringing, but others in the church as well as some laymen opposed the innovation for a variety of reasons. There was the religious objection that if God wanted to strike a building, it was presumptuous of man to interfere. (Others counterargued that since God made His laws known to man, it was no more presumptuous to apply them for our protection than it would be to cure a disease.) One leading "electrician" of the time, who had earlier attacked Franklin's theory of electricity as contrary to his own, claimed that the rods were more likely to attract lightning to a building than to preserve the building from a strike. The controversy was particularly heated in Boston where the Reverend Thomas Price warned that electricity, transferred from clouds to earth via metal rods, would concentrate in the ground. The city of Boston, where many rods were installed, would accumulate a local pool of electricity which could enhance the likelihood of earthquakes (Dibner, 1977; Cohen, 1952).

No doubt some lives and property were lost as a result of critics' successes in delaying installation of lightning rods. In that sense society would have done better without the critics. But in other instances, we have lost from the *absence* of critics. Consider another case where we again have the benefit of hindsight.

1

Up to the present day there have been numerous cases of unexplained sudden death among infants and young children. Since at least the seventeenth century such deaths have often been associated with abnormal enlargement of the thymus gland. During the first decades of this century, many physicians thought that an abnormally enlarged thymus, pressing against the windpipe, caused coughing, choking, and wheezing and, in acute cases, sudden death. This pathological condition was variously called "enlarged thymus," "thymic asthma," or "status lymphaticus." Modern medical opinion now holds that "enlarged" thymus is innocuous and not to be treated (Goldstein and Mackay, 1969).

In 1905, only ten years after x-rays had been discovered by the German scientist Roentgen, a Cincinnati pediatrician named Alfred Friedlander used the rays to shrink the thymus of an infant who he feared was in mortal danger from status lymphaticus. Friedlander knew that x-rays reduced thymus tissue in animals, so with parental consent he x-rayed the baby several times over a period of one month, and soon afterward the young patient's breathing difficulties disappeared (Friedlander, 1907).

The treatment caught on quickly, first among Friedlander's colleagues in Cincinnati and then more widely in the United States and Europe. By the 1920's many hospitals in this country routinely took x-ray pictures of all newborns to screen for enlarged thymus, and thousands of children received radiation therapy for the condition (Hudson, 1935). "Prophylactic" irradiation was given even to symptom-free infants (Conti and Patton, 1948).

All the while that status lymphaticus was the popular childhood affliction, and radiation the therapy of choice, a number of physicians doubted that the thymus could cause all the trouble attributed to it. Indeed, as early as 1858 one anatomist insisted that the soft, spongy thymus could not place serious pressure on the windpipe and denied the existence of thymic asthma. Critics multiplied in the 1920's and 1930's. Their focus was not on the potential dangers of x-ray therapy, but on whether or not enlarged thymus was indeed a cause of sudden death, and whether status lymphaticus was a meaningful diagnosis:

It has recently become the fashion [for physicians] . . . to attribute all the disturbances of infancy and early childhood, which they cannot lay to rickets, to the thymus. . . . A Roentgenogram of the chest shows, of course, the shadow of the thymus. If this is larger than they think it ought to be, it proves to them that an enlarged thymus is the cause of the symptoms. If the shadow is no larger than they think it should be, they say that there is something wrong with the picture and still attribute the symptoms to the thymus.

Whatever the size of the shadow, they are likely to advise treatment with the Roentgen ray. If the symptoms diminish or disappear at any time after treatment with the Roentgen ray, they are satisfied that the improvement was due to shrinkage of the thymus, *post hoc, propter hoc* always being a satisfactory explanation to many minds. If there is no improvement in the symptoms, they are likely to recommend more treatment with the Roentgen ray or shift to ultraviolet ray treatment or cod liver oil, although occasionally someone admits that he is wrong and looks for some other cause for the symptoms (Morse, 1928: 1547).

Eventually, improved anatomical studies showed that the thymus in normal infants is larger than previously had been supposed, that the presence of respiratory symptoms was not well correlated with size of the thymus and that symptoms usually cleared as quickly with or without radiation (Steward, 1925; Morse, 1928; Young and Turnbull, 1931; Boyd, 1932). As a result of this criticism, "enlarged" thymus was dropped as a cause for pediatric concern, and children were no longer irradiated for this condition. How could there have been so much enthusiasm for treating a disease that wasn't there?

Proponents of thymus radiotherapy usually held a set of three interlocking beliefs. First, they were quite sure that there was a real need for the therapy—that hyperthymus was a serious and common problem. Second, they were sure that radiation was an effective therapy. Third, they were fully convinced that the radiation was safe, usually on an absolute level and certainly relative to other forms of treatment that were available. These three beliefs are well illustrated by the quotes in Exhibit 1 extracted from papers by Friedlander and his associate Lange, and by the Philadelphia radiologist, G. Pfahler, who joined the proponents somewhat later. Not everyone who favored radio therapy held these beliefs so strongly, but some accepted them to an incredible degree. Two radiologists claimed "abnormally enlarged thymus is common in the new-born, occurring in from 40 to 50 percent of such infants" (Peterson and Miller, 1924: 238); in other words, it was normal to be abnormal. The x-ray was supposed to produce a curative effect by shrinking the thymus, thus relieving pressure on neighboring organs, though one enthusiast could see the effect even in the absence of the cause: "Certain types of thymic enlargement do not decrease in size after irradiation, but the symptoms in [even] these cases subside" (Donaldson, 1930: 532).

The confident optimism of the proponents blinded them to the possibility that radiotherapy might be dangerous to their young patients. Pfahler, in discussing a series of 34 irradiated patients, passed

EXHIBIT 1

Beliefs of Three Proponents of Radiotherapy for Enlarged Thymus

| Proponent | Belief in the Need for the Therapy | Belief in the Efficacy of the Therapy | Belief in the Safety of the Therapy |
|---|---|---|---|
| Alfred Friedlander, pediatrician who first used x-rays to treat enlarged thymus | [I]t appears to be indubitable that . . . the pressure of the enlarged thymus exercises a pernicious influence and does produce some of the cardinal symptoms (1907: 494). [W]e have life-threatening symptoms produced directly as a result of the enlarged thymus (1911: 811). | [T]reatment of an enlarged thymus with the x-ray was astonishingly successful (1907: 496). The value of the x-ray in cases of status lymphaticus with enlarged thymus may now be considered as established (1911: 812). | Certainly the x-ray is far safer than the dangerous operation of thymectomy. Apparently, it is not followed by subsequent ill effects, either (1911: 828). |
| Sidney Lange, a radiologist who worked with Friedlander | In the last few years abundant verification of the direct relation of enlarged thymus and sudden death is to be found in the current medical literature and the dead room reports of any large hospital (1913: 74). | In every recorded case the application of the x-ray [to an enlarged thymus] was followed by prompt and complete recovery (1913: 77). | X-ray exposure of the thymus gland had proven to be harmless. . . . (1913: 80). |
| Dr. G. E. Pfahler, later proponent of the treatment | [T]he pressure effect [of an enlarged thymus] has come to be pretty generally accepted as the cause of the local symptoms. . . . Hedinger in autopsies in 18 cases of "thymus deaths" in infants demonstrated pressure in each case (1924: 41). | [I]f symptoms do not show definite improvement after a few x-ray treatments . . . the diagnosis [of enlarged thymus] is almost certainly incorrect (1924: 39). There is probably nothing in radiotherapy that gives such uniformly brilliant results (1924: 44). | I am sure my colleagues will agree that no harm has resulted from the treatment (1924: 44). |

quickly over the fact that 2 had died, one "from inexplicable convul-sions" (1924: 45), with no thought that the x-ray could have been at fault. Barnes studied 67 children who received radiotherapy of the thymus, and after eliminating from his statistics 4 who died, con-cluded that the dosages "are apparently within the range of safety" (1929: 225). Thus, when status lymphaticus was discredited as a pathological condition, and radiation diminished as a treatment for enlarged thymus, there was still no widespread concern that thera-peutic x-rays, "properly" administered, could be dangerous.

During the 1920's, as the medical basis for thymus irradiation was destroyed, the x-ray was increasingly turned on other types of lymphoid tissue, particularly the tonsils and adenoids. Radiation caused tonsils to shrink, and as a result cavities within them drained, thus lessening infection. The shrinkage of an enlarged thymus had been considered a lifesaving technique, but shrinking tonsils was simply a matter of curing persistent sore throats. Radiologists consid-ered the x-ray so safe that it could be used even on this innocuous condition. James Murphy and W. D. Witherbee, the first American physicians to irradiate tonsils, claimed that "with the mildness of the roentgen-ray treatment recommended, there is no reason why it should not be repeated as often as desired . . . .The actual amount of roentgen ray used is smaller than that commonly used in the treat-ment of ringworm of the scalp, from which no bad results have been recorded" (1921: 228).

The x-ray was a time bomb with a slow clock. In 1910, it ticked too softly for anyone to recognize the full danger; by 1930, the radiolo-gists were so accustomed to the ticking that they did not worry much about it. As treatments for thymus declined, those for tonsils, ade-noids, and even acne increased. The number of children irradiated for these innocuous conditions has been estimated at between several hundred thousand and a million (Remsberg and Remsberg, 1976: 153).

In 1950, Duffy and Fitzgerald reported that in their series of 28 children with thyroid cancer, 10 had previously received radiation for enlarged thymus. In 1955, Clark reported 15 cases of thyroid cancer in children; all had been x-rayed for thymus or tonsils. Also in 1955, Simpson and his associates reported a significantly higher rate of thy-roid cancer in 1,400 children irradiated for enlarged thymus than in their untreated siblings or in the general population. Hempelmann (1968) found thyroid cancers appearing 10 to 20 years after x-ray therapy. DeGroot and Paloyan (1973) found thyroid tumors appearing an average of 20 years after exposure to x-ray therapy "with alarm-

6 / *The Dynamics of Technical Controversy*

ing frequency." In Chicago, where hospitals began a callback of patients who had been irradiated since 1938, clinics found three to seven percent had malignant tumors and many more had nonmalignant anomolies of the thyroid (Refetoff, et al., 1975). If the incidence of cancer nationally is similar to that found in Chicago, then x-ray treatment for innocuous conditions may have produced or will ultimately produce thousands of cancers.

There had never been vociferous opposition to x-ray treatment of children. Even the critics of status lymphaticus did not object to radiotherapy per se. *If* there had been vocal critics, could they have alleviated this technological tragedy? Of course we will never know, and each of us must be content with his own guess. Perhaps because my tonsils were irradiated when I was a child and a cancerous thyroid was removed 28 years later, I believe that critics would have provided a beneficial barrier against the proponents. A critic is not brighter or wiser than a proponent, but he usually operates by stacking up as many objections as are feasible. Invariably some of these are groundless, but they force both sides to examine carefully those assumptions which might otherwise be taken for granted, and in the process valid problems are sometimes brought to light.

In the absence of real critics we may make up a hypothetical critic and consider what he would have said at various times in the history of radiation therapy. We can be quite sure that he would have argued that "enlarged" thymus was not dangerous (since many people did in reality take that position) and therefore ought not to be treated with radiation, and in later years he would have made the same argument about infected tonsils. From hindsight we know these are valid points. In searching for issues he probably would have claimed that radiation of the thymus causes Mongolian idiocy, a concern that really was raised at one point (Hess, 1927) but never pursued vigorously. He surely would have warned about the possibility of burns and electric shocks from improperly operated x-ray equipment, a point which proponents often raised to insure that treatments were given only by qualified radiologists. But would the critic have guessed that radiation could cause cancer many years after the treatment?

Surely the cry about cancer would have been raised by 1950 when Duffy and Fitzgerald explicitly suggested that thyroid malignancies in children might be caused by radiotherapy. A similar suggestion appeared in the literature a year earlier but without supporting evidence (Quimby and Werner, 1949). In 1946, Ulrich showed that leukemia was eight times more likely to be the cause of death for radiologists than for other physicians, and he cited 15 references, dating back to 1924, which suggested that radiation causes cancer. Carman and Miller, writing in 1924, claimed "there have been a hundred victims of cancer among radiologists . . . .Studies . . . have been made of irradiation effects on virtually every organ and tissue of the body . . . .(I)t is impressive to note that alterations of some sort were found almost constantly, and that in many instances they were produced by moderate or even minute amounts of irradiation" (p. 408). Also in 1924, Brehm specifically warned of "dense tissue formation" in the thyroid setting in gradually, long after irradiation." Of course, the evidence was scant and perhaps faulty, and proponents saw no reason for concern.

Society took a loss from the critics of lightning rods, but perhaps we would have gained if there had been critics of radiotherapy. Too much opposition would have deprived us of the real benefits of the x-ray, so we are left with the obvious question, How much opposition is a good thing? The problem is to balance the proponents of a technology against its opponents so that each probes the other's position, exposing weak spots which might otherwise escape attention, but so that neither side wholly overwhelms the other. Perhaps it is naive to believe that a delicate balance is realizable, but it seems foolish to reject the notion outright, considering that we barely understand the nature of technical controversy—of the social processes and mechanisms that make up a dispute.

My intent here is to improve our understanding of technical controversies. I have tried to approach them as social phenomena which may be studied by the traditional methods of empirical sociology, so that we may understand something about the recruitment and motivations of partisans, and the formation of issues, and perhaps even predict the course of events in a mild way. I do not hold a strong position for or against any of the technologies which will be discussed in subsequent chapters, so I believe that I have avoided the common bias of regarding one side as good guys and the other as bad guys. There are other biases that are reflected in this work, however, and these include the beliefs that technical controversies can be functional for our society, that they ought to be handled better than they are currently, and that in order to handle them better we need to understand them more than we do.

It is traditional in beginning a study such as this one to define the main object of concern. Technical controversies are like any other sort of controversy in most ways; however, a few features will set them off as a convenient subset. First, a focal point of the dispute must be some product or process of science or technology (though these controversies invariably have political, religious, or other ideological issues, too). Second, some (not all) of the principal participants in the controversy must qualify as expert technologists or scientists. Third, there must be experts on opposing sides of the controversy who disagree over relevant scientific arguments which are too complex for most laymen to follow. These criteria do fairly well in specifying those controversies which most observers regard as "technical" though there will be occasional ambiguous cases.

The remainder of the book is implicitly organized into three parts plus a concluding chapter. The first of these, consisting of Chapters 2 and 3, will discuss the scientific disputes between experts which are a part of every technical controversy. At times the resolution of a controversy seems to hinge on the outcome of a scientific argument, and at other times the controversy appears to proceed quite independently of these technical issues. In all cases, however, a great deal of attention is given to the complex arguments of experts which bewilder laymen trying to understand the issues. Chapter 2, "Disputes Between Experts," compares technical arguments found in the controversies over fluoridation and radiation, showing similarities in both the structure of the arguments and the behavior of the experts. In Chapter 3, "Separating Disputes Over Facts from Disputes Over Values," I explore the practicality of simplifying these arguments by treating contentious scientific issues apart from the nonscientific issues with which they are usually intermeshed.

The second part of the book examines the principal actors in a technical controversy, the scientists and nonscientists who promote or attack a technology. In Chapter 4, "Partisans," I compare the active proponents and opponents to each other, and also to the great majority of the public which takes no active role beyond registering their opinion for or against a technology in an attitude poll or on a referendum. Chapter 5, "Beliefs, Ideology, and Rhetoric," is concerned with the motives, beliefs, and ideologies of partisans, and with asymmetries in the controversies which encourage the development of particular forms of rhetoric and issue content. In Chapter 6, "Social Links Among Controversies," I trace the movement of academic scientists across a series of political controversies, nontechnical as well as technical ones, in order to explicate how social networks among partisans may link together a number of separate controversies.

I switch my focus from individual partisans to the protest itself in the third part of the book. Why do some technologies become sources of widespread public protest, even spawning mass movements as in the cases of nuclear power and fluoridation, while other technologies, no less hazardous, are tolerated? Chapter 7, "Growth of Protest," attempts to answer this difficult question by describing steps taken by these movements in their expansion. Even the very largest protests show marked variations in intensity after they have achieved the status of full-blown mass movements, and the causes of these fluctuations are the subject of Chapter 8, "Rise and Fall of Controversy."

It seems to me that technical controversies can be extremely useful for society by providing an effective means for identifying and evaluating the problems and advantages of technologies which are not always in clear view. I will argue this position in the final chapter and suggest ways in which this function might be performed better than it is now.

# 2
# Disputes Between Experts

We may define "experts" as two or more people who can authoritatively disagree with one another. Virtually any case study of a technical controversy will contain references to each side's experts—the properly credentialed scientists, engineers, and physicians who buttress its positions with technical information and who undermine the scientific basis of the other side. Indeed, I have taken the existence of a dispute between experts as a defining characteristic of technical controversies.

Laymen are often confused and dismayed when one scientist contradicts another's facts. There is a popular conception that scientists know the truth—at least in their domains of inquiry—and if two of them disagree, then one must have lied or made a mistake. Of course, science is not equivalent to truth, and there are ample opportunities for two competent scientists to disagree with one another, particularly on issues at the state of the art. Just as historians used to chart the course of empires by tracing the links from one war to another, one could write a passable history of modern science by linking the great theoretical and experimental controversies. Scientific disputes which seem silly in retrospect apparently were taken quite seriously in their time. I can think of no better example than the controversy over the efficacy of prayers for the sick.

This chapter is adapted from an article by the author, entitled "Disputes Between Experts," which originally appeared in *Minerva*, XI, 2, April 1973, pp. 243-62; and includes excerpts from an article by the author with Andrew Marino and Robert Becker, entitled "Separating Factual Disputes from Value Disputes in Controversies over Technology," which originally appeared in *Technology in Society*, Vol. 1, 1979, pp. 229-37. Used with permission.

The setting was Victorian England in a period of growing prominence for science and increasing conflict with organized religion. Sir Henry Thompson, a London surgeon, published an article which argued for an experimental test of the effect of prayers for the sick. Sir Henry suggested that the efficacy of prayer could be tested in the same way we test any other proposed remedy. Select a hospital with a well established mortality rate and, for a period of at least three years, make the patients of that hospital "the object of special prayer by the whole body of the faithful." At the end of this period, the hospital's mortality rate would be compared with its prior mortality and also with rates of similar "control" hospitals where patients were not the special objects of prayer. If prayers for the sick were indeed effective, a decline in mortality would be observed in the "treatment" hospital.

The theologians were quick to counterattack, pointing out that God could hardly be expected to cooperate with this arrogant experiment, and therefore, a null finding would be uninterpretable. Francis Galton entered the fray with some relevant statistical data he had collected over the past several years. Since British subjects frequently shout "God save the King," Galton reasoned that British monarchs should have greater longevity than lawyers, gentry, and military officers, but in fact the opposite was true. Furthermore, clergymen, who pray frequently, do not live significantly longer than lawyers and physicians; and missionaries often die shortly after arrival in the foreign land where they were to spread the gospel, in spite of the numerous prayers that accompany them. The theologians then cited instances where large-scale prayer *had* been effective, such as for the spread of Christianity and the longevity of the Papacy. Reverend James McCosh, president of Princeton College in America, pointed out that God may answer a prayer for the sick in some way other than curing the patient. For example, when Prince Albert was sick a few years before, thousands prayed for his recovery, but he died. Shortly afterward, when Queen Victoria's advisors urged her to go to war with America, the Queen refused on the grounds that her late husband opposed warfare between England and America. McCosh argued that Prince Albert's death made his wishes particularly influential on the Queen. So in refusing to answer the thousands of prayers for Albert's recovery, God was really acting in the best interests of those who prayed (Brush, 1974).

The debate raged in the journals during the years 1872-73 and then dropped from sight, apparently without producing any major effect on the popular conception of prayer. We may wonder what would have happened if the hospital experiment had been carried out and mortality declined in the treatment hospital. In any case, the prayer controversy illustrates a number of features which are observable in contemporary disputes between experts. Hypothetical experiments are proposed and attacked for polemic effect. Experimental results are discredited or reinterpreted to suit either side. Data are used selectively to support one position or another.

Experts frequently disagree on scientific and technological questions which are relevant to political issues. Some of these questions have been labelled "trans-scientific" by Alvin Weinberg (1972) because they are in principle beyond the capacity of science to answer. For example, it would require so many mice, so much time, so many scientists, and so much equipment to obtain significant results on the biological effect of very low-level radiation that the experiments would probably never be undertaken. Other questions can be answered but for one reason or another have not been. In either case, such disagreements between scientists who testify as experts is a major source of confusion to policy makers and to the public.

One example is the disagreement over the harmful effect of low-level radiation. This dispute originated in early concern over the fall-out from nuclear tests and persists in the current controversy over nuclear power plants (Marx, 1979). Major critics in the late 1960's were Dr. John Gofman and Dr. Arthur Tamplin, research associates at the Lawrence Radiation Laboratory. Gofman is also professor of medical physics at the University of California, Berkeley, and a past associate director of Lawrence. In late 1969, Gofman and Tamplin claimed that if the population of the United States were exposed to the maximum level of radiation permitted by federal standards, there would be an additional 16,000 to 32,000 cases of cancer and leukemia each year. They recommended a ten-fold reduction in the federal standards. The scientific reception of their work has been well described in *Science*: " . . . the Atomic Energy Commission [AEC] challenged their assumptions, disputed their estimates, and disagreed with their recommendation" (Boffey, 1970). Still, while many scientists who are experts in the subject disagreed with them, the two are generally considered to be reputable scientists whose arguments are not clearly wrong. Another *Science* article noted that: "Most scientists who have worked on setting [radiation] standards believe that many of the assumptions made by Gofman and Tamplin are unjustifiable but find it difficult to disprove specific points" (Holcomb, 1970).

### Nuclear Power and the Fluoridation of Water

In order to clarify my own confusion on this sort of technical disagreement, I was led at that time to study the conduct of opposed experts, comparing the nuclear power controversy with the water fluoridation controversy which began in the 1950's. Both controversies focused in large part on similar technical questions: what are the harmful effects, if any, of long-term exposure to low-level doses of fluorine or radiation? The two questions have similar patterns: fluorine and radiation are known to be lethal in large doses and although there is no clear evidence of their lethal effects in very low doses, neither—so some experts argue—is there compelling evidence to the contrary.

There is a popular stereotype of the antifluoridationist as a "kook," bigot, and extreme "right-winger." While some opponents of fluoridation could have been described in these terms, it is necessary to recognize that some respectable scientists and physicians have also opposed fluoridation, fearing possible toxic effects. Yet few "neutral" commentators have given serious consideration to their arguments. Two psychologists have called opposition to fluoridation an "antiscientific attitude" (Mausner and Mausner, 1955). Social scientists have been inclined by and large to assume that an informed voter could not rationally oppose fluoridation, and they studied its frequent defeat in referenda as examples of "democracy gone astray."[1] In comparison, the critics of radiation levels were given a respectful—if not hospitable—reception.

There are four plausible explanations for this apparent difference in the treatment of scientific opposition. First is the possibility that the radiation argument is objectively more sound than the fluoridation argument. Second, the antiradiation scientists might have higher professional stature than the scientists who oppose fluoridation. Third, the antifluoridation movement was associated with the anticommunist campaign of the late Senator McCarthy, and that has been anathema to the American academic and scientific communities. Fourth, it is only in the last few years that scientists, and the public, have become acutely aware of, and concerned with, "traces" of mercury, DDT, etc. in the environment. It is ironic to read the facetious discussion of Crain and his colleagues of claims against fluoridation "on alleged medical grounds." This passage was published in 1969 in a sociological survey of attitudes towards fluoridation, just as the Environmental Movement was beginning to sweep the United States:

[Most of the] . . . claims made against fluoridation on alleged medical grounds . . . have their basis in the fact that in concentrated dosage fluorine is a poison. When the proponents of fluoridation try to argue that one part per million is a highly diluted dose, the critics reply that the fluoride will collect in out-of-the-way corners of the water mains to build up to deadly dosages. The reputed side effects of fluoridation run from destruction of teeth to liver and kidney trouble, miscarriages, the birth of mongoloid children, and psychological disturbances, including suspectibility to communism and nymphomania. When the public-health officer points out that nearly a tenth of the drinking water in the United States has always had traces of fluoride in it without causing ill effect, the critics then charge that fluoridation damages car batteries, rots garden hoses, and kills grass (Crain, et al., 1969:4).

Some of these arguments do not sound quite as nonsensical today when many persons worry about a "highly diluted dose" of mercury or cyclamates. Some radionuclides, mercury, and DDT "build up" through the now well-known ecological process of chain-concentration in food—fluorides concentrate in fish and tea. And the public health officer's argument that some water "has always had traces of fluoride in it without causing ill effect," sounds very much like the current pronuclear argument that we have always been exposed to background radiation without injurious effects.[2]

In most of what follows, I will examine the scientific or technological content of these disputes, but comparative analysis suggests that the political, nonscientific context of the dispute—e.g., McCarthyism or "environmentalism"—might be equally important in determining the outcome.[3]

## Rhetorical Devices in Technical Disagreements

Before examining the technical similarities in the disputes over radiation and fluoridation, it is useful to look briefly at the rhetorical similarities in the technical controversies as they have appeared in periodical articles, speeches, congressional hearings, and reports in the press. Perhaps the rhetorical devices, more than conflicting substantive arguments, are the main source of public confusion. Even a casual reading of the literatures opposed to fluoridation and nuclear power reveals their similarity. In what follows, I present several passages from Gofman and Tamplin, opposing nuclear power because of the radiation hazard, each followed by a similar passage from an opponent of fluoridation.

*Radiation:* The freshwater-to-fish pathway can concentrate radioactivity easily 1,000-fold or more . . . . Thus, even though the water effluent at the release point may make the water drinkable . . . the fish grown in such water, 1,000 times as radioactive, cannot be eaten in any quantity without grossly exceeding "tolerance levels" (Gofman and Tamplin, 1971:307).

*Fluoridation:* People living in fluoridated cities who eat a good deal of seafood and drink tea and beer may easily ingest a combined fluoride intake far beyond even the tolerance limits assumed by the Public Health Service (Exner, et al., 1957: 20).

*Radiation:* . . . The AEC clearly demonstrated that when the chips are down on questions of protecting human beings and their environment, the promotional huckster role wins out handily over the public protector role (Tamplin and Gofman, 1970:123).

*Fluoridation:* . . . the reckless arrogance, obstinancy, and unscrupulousness of the United States Public Health Service in continuing to promote the program while ignoring and, where possible, suppressing evidence that it is neither safe nor genuinely efficacious (Exner, et al, 1957:12-13).

*Radiation:* Where unknowns exist [in the evaluation of a technology], always err on the side of protecting the public health (Gofman and Tamplin, 1971:257).

*Fluoridation:* . . . the public should have the benefit of the doubt and the procedure should be considered harmful until proved otherwise (Taylor in *Hearings,* 1952:1535).

*Radiation:* Where environmental poisons are concerned, it has always been up to the public to show harm, rather than up to the pollutor to prove safety. The promoters of atomic energy . . . said, in effect, the public must prove it is being harmed by radioactivity . . . (Gofman and Tamplin, 1971:246, 257).

*Fluoridation:* When a potentially dangerous substance such as fluoride is added to a public water supply, the burden should rest on those who add it to prove beyond reasonable doubt that it is safe for everyone. This has not been done. In fact, there is a strong reverse tendency to require incontrovertible proof of damage from opponents . . . (Exner, et al., 1957:45).

*Radiation:* Tamplin and Gofman presented evidence . . . that our allowable radiation exposures . . . are grossly unsafe . . . . The AEC response: Derision, denial, slander—but no evidence in refutation (Tamplin and Gofman, 1970:223).

*Fluoridation:* [Critiques] of the proponent scientific data have been presented to the Public Health Service . . . . Instead of dealing with the subject matter itself, they [the PHS] attempt to show that the author is not qualified to discuss matters related to fluoridation (Exner, et al., 1957:188).

*Radiation (referring to the strategy of the AEC):* Tell a big lie and tell it again and again as widely as possible (Tamplin and Gofman, 1970:123).

*Fluoridation (referring to the strategy of the PHS):* . . . a colossal lie if repeated often enough, will be accepted as truer than truth (Exner, et al., 1957:145).

Lest this similarity be considered a unique characteristic of the critics, here are comparable passages from the proponents of nuclear power and fluoridation.

*Radiation:* . . . radiation is by far the best understood environmental hazard (*Electric Power and the Atom,* undated:12).

*Fluoridation:* . . . never before has a public-health measure been subjected to such thorough scientific scrutiny . . . (Forsyth in *Hearings,* 1952:1484).

*Radiation:* It seems that a number of national concerns have converged to make up what we call the nuclear controversy . . . . [One of these is] an increasing distrust of science and technology in general (Slater, 1970:2).

*Fluoridation:* The strength of the opposition to fluoridation can be attributed to three important factors . . . . [One of these is the public's] current suspicion of scientists (Mausner and Mausner, 1955:39).

*Radiation:* It must indeed be confusing to the public to have two scientists present such opposing views, and the important question arises as to which to believe. In making up your mind, I believe it important that you consider the views of the majority of scientists on these issues (Bond, 1970:8).

*Fluoridation:* The issue of fluoridation . . . came down to a question of . . . what authority . . . [the public is] to trust—the professional organizations [which supported it] or the few individual doctors, dentists [and] scientists . . . [who] opposed it (Hutchinson in McNeil, 1957:171).

*Radiation:* . . . the risk of nuclear power is very much lower than the risk of alternate power sources . . . . Compared to the benefits of electricity . . . nuclear power is a very satisfactory system (Starr, 1971:40).

*Fluoridation:* . . . the risk that such patients [with chronic kidney disease] might be harmed by the fluoridation of water appears to be small in comparison with the dental benefits to be obtained [for the community] (Heyroth in *Hearings,* 1952:1504).

*Radiation:* Part of the rationale behind permitting the release of small quantities of radioactivity to the environment is the knowledge that the environment has been radioactive from natural causes since the beginning of time. All natural solids, liquids, and gases contain radioactivity in varying amounts. Further, radiation due to cosmic rays continuously bombards us (Seaborg and Corliss, 1971:70).

*Fluoridation:* We have analyzed foods very common to our diet which were purchased on the open market and have found that they contain fluorine in amounts varying from 0.14 to 11.2 parts per million. Therefore, the addition of fluorine at approximately 1.0 part per million to the water is not introducing a new element into our dietary (Blayney in *Hearings,* 1952:1548).

*Radiation:* After more than ten years of experience with nuclear power, no utility-operated nuclear station in this country has ever had an accident that adversely affected public health (Electric Light and Power Companies, 1971).

*Fluoridation:* In the more than 200 municipalities that have fluoridated their water supplies, no serious problems have occurred (Doty in *Hearings* 1952:1678).

Many of these statements could be transposed from one controversy to the other simply by changing "radiation" to "fluoridation," and "AEC" to "PHS," or vice versa.

The most common rhetorical device appears to be the phrase, "There is no evidence to show that . . . " or one of its variants. This blank denial of the claims of the opponent on the ground that there is no basis for his position appears in both controversies.


*Fluoridation:*

No evidence has ever been produced that 1.0 part per million of fluoride in drinking water has or will harm any living person or thing (Forsyth in *Hearings,* 1952:1485).

I would say that as far as any evidence has brought out to the present time, there is no danger to our health and welfare (Blarney in *Hearings,* 1952:1559).

Some surveys of the amount of certain types of kidney disease in fluoride as compared to nonfluoride areas have not produced any evidence of harmful effects upon the kidneys by fluorine at the levels proposed for the fluoridation procedure (Doty in *Hearings,* 1952:1678).

The councils [of the American Medical Association] are unaware of any evidence that fluoridation of community water supplies up to a concentration of one part per million would lead to structural changes in the bones or to an increase in the incidence of fractures (Lull in *Hearings,* 1952:1709).

In the accumulated experience there is no evidence that the prolonged ingestion of drinking water with a mean concentration of fluorides below the level of causing mottled enamel would have adverse physiological effects (Heyroth in *Hearings,* 1952:1504).


*Nuclear Power:*

No evidence exists for such an effect [i.e., differential harm depending on the rate of radiation delivery] on cancer or leukemia induction by radiation in man (Gofman and Tamplin, 1969, GT-102-69:14).

At the present time no valid evidence, based upon scientific observation, has been brought forward to prove that natural sources of radiation have produced injury to man in any way (Gofman, quoted in Bond, 1970:3).

There are no experiments that show that the integrated low-level effect [of radiation] is higher than that of the same amount given at one time (Starr, 1971:38).

The device occurs on both sides of a single controversy. Thus, the first two "Nuclear Power" statements above are by Gofman, the first made from his recent role as a nuclear critic, and the second 12 years earlier when he was a proponent. Not surprisingly, there are counter-moves to the "no evidence" rhetoric:

Witness a typical statement by Mr. Frederick Draeger of the Pacific Gas and Electric Company: "There is no evidence that 170 millirads is harmful and any new plant will actually emit only an infinitesimal fraction of that amount." Apparently, Mr. Draeger hasn't the slightest comprehension of what his statement "no evidence" really means. "No evidence" here means no one has even looked (Gofman and Tamplin, 1971:104-5).

We find the same mode of argument in the fluoridation debate. For example, when the American Medical Association stated that its responsible councils "are unaware of any evidence" that fluoridation would be harmful, an opponent responded:

[If] the Councils had actually considered the evidence instead of trustingly accepting what McClure said about the evidence, they would not have been unaware of dangers in fluoridation (Exner, et al., 1957:79-80).

## Arguing Past One Another

Some observers, in the course of trying to place their finger on the points of disagreement between two experts, have concluded that the two do not disagree at all, but rather are each arguing about different points. This failure to confront each other's arguments is clearly present in the dispute surrounding the Gofman-Tamplin analysis of the expected number of deaths from the nuclear power program.

Gofman and Tamplin calculated that the United States would have 32,000 cancer plus leukemia deaths annually from population exposure to the maximum radiation level permitted by federal standards. One of their opponents, Dr. V. P. Bond, of the Brookhaven National Laboratory, clearly stated the opposing view (Exhibit 2). Note that Dr. Bond shows "cancer cases per year" to be the product of three factors: "risk" (measured in cases of cancer per million population per mrem of exposure), "dose per year" (measured in mrem per year), and "number of persons" (taken as 200 million). Gofman and Bond differ in the values they assign to "risk" and "dose per year," and therefore, they arrived at markedly differed values of "cancer cases per year."

EXHIBIT 2

Comparison of Gofman and Bond Calculations

| Risk | × | Dose per Year | × | Number of Persons | = | Cancer Cases per Year |
|---|---|---|---|---|---|---|
| (cases per 1,000,000 population/mrem) | | (mrem per year) | | (millions) | | |
| Gofman's calculation 0.94 | × | 170 | × | 200 | = | 32,000 cases per year |
| Bond's calculation 0.1 | × | 0.001 | × | 200 | = | 0.02 cases per year |

Source: Bond, 1970: figure 1.

The values for "risk" differ by an order of only ten (0.94 versus 0.1), and we will pass over this difference for the present. The values for "dose per year" differ by an order of $10^5$ (170 versus 0.001). Gofman's value is based on the permissible (but not actually achieved) level of exposure. Bond's value is his estimate of actual average exposure to the population, which is much smaller than the permissible exposure. The two calculations are about two different things. Bond concludes:

Dr. Gofman's speculations that 32,000 additional cancer deaths per year will result from radiation exposure of the public under current "standards" simply do not conform to reality. They are in fact *in error* by a considerable margin for the present and for the foreseeable future. His figures have essentially zero validity in the context of power reactors. In this context, an upper limit estimate of the *correct* figure is well below one death per year in the entire U.S.A. (Bond, 1970:12, italics added).

Of course, the question of what is "in error" and what is "correct" depends on what is being calculated.

In both controversies, there are similar patterns of conflicting contentions based on differing premises, one of these involving the difference between acute and chronic forms of radiation or fluoridation poisoning. Proponents of fluoridation and nuclear power have occasionally argued for the safety of their proposed technology by indicating how difficult it would be for a person to receive the relatively high dose associated with acute poisoning:

[Even] at one part of fluoride per million parts of water, to get a lethal dose from it you would have to drink 400 gallons at one sitting (Forsyth in *Hearings*, 1952:1504).

The critics, on the other hand, were concerned about chronic poisoning which is associated with much lower dosages:

All the talk about the hundreds of gallons [of water] you would have to drink at one time to get sick refers to acute [fluoride] poisoning, which isn't even under consideration. . . . What is important is that the presence of tiny amounts of fluoride in the tissue fluids for long periods interferes with the proper growth, development and function of many parts of the body (Exner, et al., 1957:37).

These conflicts based on divergent premises appear to result from poor communication between adversaries, and/or from a strong motivation to win the argument. As such, they could probably be eliminated from a public debate by qualified persons so that the technical issues would stand out more clearly. There is another source of confusion, however, which appears at the heart of technical disagreements and which could probably not be eliminated from any debate because it is intrinsic to disagreements. Even with perfect communication and eschewing rhetorical devices which are intended simply to put the opposing argument in an unfavorable light, experts may disagree on ambiguous observations and assumptions which cannot be resolved by available objective means.

## Ambiguities

The theories, models, procedures, and formulas of science and technology are generally believed to allow one trained in their use simply to calculate an unambiguously correct answer. A technologist or scientist soon comes to recognize that the complex technical problems of the state-of-the-art require subtle perceptions of the sort which cannot be easily articulated in explicit form. When it is necessary to make a simplifying assumption, and many are reasonable, which simplifying assumption should be made? When data are lacking on a question, how far may one reasonably extrapolate from data of other sources? How trustworthy is a set of empirical observations? These questions all require judgements for which there are no formalized guides and it is here that experts frequently disagree. I will call these points of disagreement "ambiguities," and I will demonstrate how they enter into technical controversy.

Most experts agree that radiation increases the incidence of leukemia and thyroid cancer in an exposed population. However, there is disagreement on whether other forms of cancer are similarly induced by radiation:

Wanebo et al. . . . have recently reported that "accumulated information . . . strongly suggests that exposure to ionizing radiation has increased the risk of lung cancer among atomic bomb survivors." These investigators observed 17 such cases, as compared with 9 expected. . . . [They also] have reported that "information on breast cancer . . . has now accumulated to the point where a fairly definite carcinogenic effect seems established." Six cases were observed . . . as compared with 1.53 cases expected—an excess of only 4.5 cases . . . .

It may be difficult or impossible to avoid certain biases that could produce such a small excess. . . . Wanebo et al. considered the possibility of biases and believed that none were present. . . .

*One may conclude* that . . . [the] evidence pertaining to cancer of the breast or lung *is still very much in doubt* (Miller, 1969: 572, italics added).

Consider, now, the plight of someone who is trying to calculate the number of cancers to be expected in a population exposed to a given level of radiation. Does he calculate an increase in just leukemia and thyroid cancers, or does he calculate an increase in all forms of cancer? Since leukemia and thyroid cancers constitute about ten percent of all cancers in the United States, these two calculations will differ by about a factor of ten. Gofman's and Bond's calculations (Exhibit 2) showed their "risk" values differing by about a factor of ten (0.94 versus 0.1). Gofman's calculation dealt with "all cancers" and Bond's with only leukemia and thyroid cancer. We are in no position to say which one is "correct" given the indeterminacy of the present state of knowledge. The adversaries take a less equivocal position:

Dr. Gofman's excessive estimates are based on the untenable assumptions that all forms of cancer are increased by exposure at low doses and rates. . . . These assumptions do not square with the facts (Bond, 1970:2).

And on the other side:

[Almost] all the major forms of human cancer were by [1969] . . . already known to be produced by ionizing radiation. . . . So it became possible to state a primary principle, or "law" of radiation production of cancer in humans.

That principle or law states, "All forms of human cancer are, in all probability, induced by ionizing radiation" (Tamplin and Gofman, 1970:13).

Each has chosen to accept as a firm conclusion what others regard as only tentative hypotheses. But their conclusions cannot be considered "wrong" in the sense in which an arithmetic solution can be wrong. Scientific "truths" are never proved but only gain increasing acceptance (and even then are often found to be incorrect). The point at

**EXHIBIT 3**

Comparison of Linear and Threshold Models

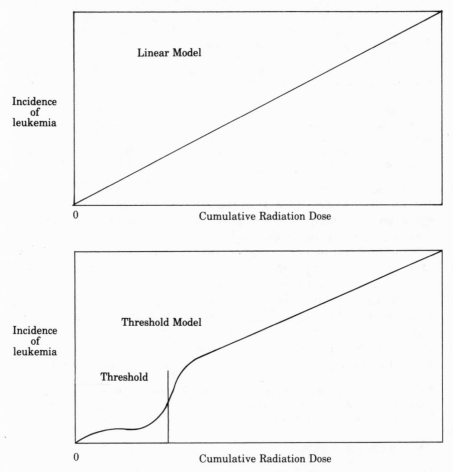

which a hypothesis becomes a conclusion differs from one scientist to another.

Given the inconclusive nature of available data, it is possible to postulate several different relationships between the radiation dose delivered to a population and the resultant increase in leukemia. Presumably, with more complete data, some of these relationships would be demonstrably inappropriate, but the data are incomplete. There are two commonly assumed "dose-effect" curves relating cumulative dose of radiation (i.e., from birth) to the population, to the incidence of leukemia in that population (measured in, say, number of cases per year per million persons). These are the "linear" and "threshold" models (Exhibit 3).

The first assumes a simple linear relationship between dose and incidence of leukemia; it is the model favored by Gofman and Tamplin. The second assumes that there is a "threshold" dose level below which there is practically no incidence of leukemia. The scanty data available are not inconsistent with either model:

There is evidence for linear dose-effect relationships of various slopes depending upon the specific effects; there is also evidence for at least practical thresholds of effects, but generally speaking there has been no statistically significant information obtained on dose-effect relationships for doses of less than a few rads, or tens of rads, delivered more or less all at once (Taylor, undated:13).

At present there is little basis for saying that one model is "true" and the other is not. It is easily conceivable that new data could prove one model wrong, but it is difficult to see how one could be "proved" correct. One could show that a given model is consistent with all available data, but it is always possible to design alternative models which fit a given data set. Thus, there is always an element of judgment in selecting one model over another empirically-consistent alternative.

This theoretical ambiguity has major implications for the technical debate over permissible radiation standards. It should be noted that the "threshold" model implies that dose levels below the threshold will not harm the population (through leukemia). The "linear" model implies that there will be some incidence of leukemia no matter how low the dose to the population. The two models differ, then, on whether or not there is a "safe" level of radiation exposure for the population. The ambiguous nature of the dose-effect curve is well recognized by radiation biologists, and many (including opponents of Gofman and Tamplin) assume the "linear" model, not necessarily because they consider it true, but because it is more conservative for purposes of public safety.

It would be reasonable, given these ambiguities, for opposing experts to "agree to disagree" and to suspend the debate, at least until new data permitted the issue to be reopened. That does not usually occur, however. Instead, each opponent tries to build his case, not necessarily for his adversary, but frequently for a third party: the public, a congressional committee, scientific peers, etc. In the ensuing dispute, adversaries often use one or both of two polemic strategies which can be conveniently applied to ambiguities. These are to reject discrepant data, or to present alternative interpretations.

## Rejection of Discrepant Data

A common way to deal with data which are inconsistent with one's own position is to deny their scientific validity. We have already seen Miller reject the Wanebo, et al. conclusions (that breast and lung cancers are radiation-induced) by noting: "It may be difficult or impossible to avoid certain biases that could produce such a small excess . . . ." (1969:572). I can elaborate by considering one of many similar examples from the fluoridation controversy.

Dr. Alfred Taylor showed experimentally that when a strain of mice normally susceptible to mammary cancer were regularly fed fluoridated water, the tumors appeared earlier than in control mice fed nonfluoridated water.[4] Taylor reported his findings to the PHS whereupon H. Andervont visited Taylor's laboratory. Andervont later testified that the experiments were not valid:

We came to the conclusion that inasmuch as the food [Taylor] was feeding to his mice contained 30 to 40 parts per million of fluoride, that the ½ part per million [fluorine] in the drinking water could not conceivably have had much influence on his results (*Hearings*, 1952:1666-67).

One could discredit Andervont's denial of validity to Taylor's analysis by pointing out that fluoride consumed in water is almost completely absorbed into the blood, whereas fluoride consumed as a solid must first be digested and smaller amounts will be taken up by the blood; therefore, the high fluoride content of the food does not necessarily overwhelm the fluoride in the water. Exner made essentially that objection and also emphasized the fact that both experimental and control groups were given the high-fluoride food, but only the experimental group received fluoridated water (Exner, et al., 1957:32-33, 71-72).

Armstrong, Bittner, and Treloar (1954:307-9) conducted an experiment to check Taylor's result. The mean age at which tumors appeared in their experimental mice (which had been given fluoridated water) was lower than that of their control mice. However, the difference between conditions was not "statistically significant" at the .05 level, and they considered the result attributable to chance. Taylor tried to deny the validity of these findings by arguing that Armstrong, et al. did not use a large enough number of mice. "A control group consisting of 31 animals would be insufficient to reveal differences of the order of those encountered in the work here." (Quoted in Exner, et al., 1957:185). But the control groups in Taylor's own experiments always contained an even smaller number of mice! The subjective nature of these attacks and rebuttals is clear.

## Alternative Interpretations

Even if both disputants in a technical argument accept the validity of a datum, the interpretation of that datum remains an ambiguous procedure. During the initial excavations for the contested nuclear plant at Bodega Bay, California, an earthquake fault was discovered running through the shaft. It was further determined that there had been no movement along the fault for about 40,000 years. Apparently, no one contested this fact, but two opposite interpretations could be made. Proponents of the plant claimed the fault was inactive, and there was little likelihood of future movement. Opponents claimed that since there had been no earthquake for a long time, one was due (Novick, 1969:42-44).

Statistical data are particularly amenable to alternative interpretations, especially if they contain substantial error variance, as is usually the case in epidemiological studies. H. Trendley Dean in 1938 analyzed the incidence of caries (cavities) in children in two sets of cities, one high in the presence of fluorine, the other low. He concluded that children using water with a higher fluoride content were

EXHIBIT 4

Fluoride Versus Incidence of Caries
(Permanent Teeth)

| Cities | Number Children Examined | Range of Fluoride in PPM | Percentage Caries-free |
|---|---|---|---|
| | | Dean's Display | |
| Pueblo, Junction City, East Moline | 114 | 0.6-1.5 | 26 |
| Monmouth, Galesburg, Colorado Springs | 122 | 1.7-2.5 | 49 |
| | | Exner's Display | |
| Pueblo | 49 | 0.6 | 37 |
| Junction City | 30 | 0.7 | 26 |
| East Moline | 35 | 1.5 | 11 |
| Monmouth | 29 | 1.7 | 55 |
| Galesburg | 39 | 1.8 | 56 |
| Colorado Springs | 54 | 2.5 | 41 |

Sources: Dean, 1938; Exner, et al., 1957.

more caries-free than those using lower fluoride water (Exhibit 4). Exner, an opponent, analyzed the same data in unaggregated form (Exhibit 4) noting:

It would appear to take some ingenuity and a certain amount of determination to deduce from these data the conclusion Dean drew (Exner, et al., 1957:114).

McClure, a proponent of fluoridation, later used the same data but again in dichotomized form (1970:81).

Adversaries in the nuclear controversy treat data regarding radiation-induced cancer in a similar way to support their own positions. Evans (1966) collected data on radium workers showing that no cancers occurred below a median dose level of .55 microcuries (Exhibit 5). Evans considered this support for the "threshold" dose-effect curve. Gofman and Tamplin, however, believe that the same data fit their "linear" dose-effect curve. Here is their reasoning.

First, they estimate the probability of finding cancer in a subject who has been exposed to a given dose of radiation. Focusing on the 5.5 microcuries median-dose group (as the largest and hence most statistically reliable), they note 14 cancers out of a total of 40 cases, so there is a 14/40 probability of cancer per person. This is for a median dose of 5.5 microcuries. The probability of cancer per person per microcurie is then $14/(40 \times 5.5) = 0.064$. Now, there are 80 cases with a median dose of .055 microcuries so, assuming the linear hypothesis, the expected number of cancers in that group is $0.064 \times .055 \times 80 = 0.28$ cases. But human cancers cannot occur in fractions, so the most likely outcome is zero cancers in this group, and that is what is found. A similar analysis for the lower median doses shows that in each group the expected number of cancers is near zero. Gofman and Tam-

EXHIBIT 5

Exposure to Radiation Versus Incidence of Cancer

| Number of Cases | Median Dose (in microcuries of Radium equivalent residual) | Number of Cancers |
|---|---|---|
| 42 | <0.001 | 0 |
| 61 | 0.0055 | 0 |
| 80 | 0.055 | 0 |
| 32 | 0.55 | 3 |
| 40 | 5.5 | 14 |
| 14 | 55. | 2 |

Source: Evans, 1966.

plin thus argue that the data are fully consistent with the linear dose-effect curve, and that the apparent threshold is due to very small groups of persons being exposed (1969, GT-103-69).

Alternative modes of interpretation are often used to explain away an opposing argument. Profluoridationists found no fluoride poisoning in cities with naturally fluoridated water (at about one ppm) and concluded that low concentrations of fluoride must be safe. Critics argued that low concentrations of fluoride would cause poisoning, but that calcium is an antidote. Since calcium usually occurs naturally in the same waters where fluoride occurs naturally, this explains the lack of poisoning (Exner, et al., 1957:101-2).

Proponents of nuclear power minimize the harmful effects of long-term low-level radiation because animal experiments have indicated that a given dose of radiation over a protracted period is less harmful than the same dose delivered in a short period of time. In a typical experiment of this sort, one group of ten-week-old mice is placed on a daily schedule of small doses. A second group of ten-week-old mice is given an acute dose equal to the integrated protracted dose of the first group. The final incidence of cancer is usually higher in the second group than the first group. Gofman and Tamplin dismiss the mitigating effects of protracted dosage by pointing out that for a given dose, and dose-rate, of radiation, more harm will be done to a younger organism than to an older organism. The mice receiving acute doses are fully irradiated at ten weeks of age, whereas the protracted group receives most of its radiation at an age beyond ten weeks. Therefore, the lower incidence of cancer and leukemia in the protracted group is simply a consequence of their being older when they were irradiated:

If most experimenters had delivered their acute radiation dose at the end of the protraction period rather than at the beginning, the literature would by now be filled with a different illusion—namely, that protracted radiation is more carcinogenic than acute radiation (1970, GT-109-70:15).

## Polarization

If an expert may reasonably take any one of several positions on a technically ambiguous point, then we should ask why some experts take one position while other experts take another—often opposing—position. One's interpretation of ambiguous data is often tied to one's position on the technology in controversy. Thus, since a "threshold"

EXHIBIT 6

Chronology of Gofman and Tamplin Polarization

| Date | Estimate of Annual Harm | Recommended Change in Policy |
|---|---|---|
| Oct. 29, 1969 | 16,000 additional cancer plus leukemia cases | Reduce FRC guideline exposure by a factor of 10 |
| Nov. 18, 1969 | 16,000 | Factor of 10 |
| Jan. 28, 1970 | Above 16,000; nearer to 32,000 or even higher | Factor of 10 |
| Feb. 9, 1970 | 16,000 cancer plus leukemia cases | Not mentioned |
| Feb. 20, 1970 | 32,000 cancer plus leukemia cases | Not mentioned |
| Mar. 30, 1970 | Not mentioned | Specific reductions for radiation workers |
| Apr. 7, 1970 | 32,000 | Guideline exposure should be zero, and the privilege of releasing radiation must be negotiated |
| Apr. 22, 1970 | 32,000 cancer and leukemia cases plus a large number of genetic deaths, plus a large number of deaths from other causes | Zero release |
| June 29, 1970 | 32,000 cancer plus leukemia cases, plus increases in genetically based diseases | Not mentioned |
| Aug. 20, 1970 | 32,000 cancer and leukemia cases, 150,000 to 1,500,000 genetic deaths, plus a 5 percent to 50 percent increase in such diseases as schizophrenia and rheumatoid arthritis | Five-year moratorium on aboveground nuclear power plants, and also an injunction against fast breeder reactors for an indefinite period |
| *ca.* Sept., 1970 | Same as above | Zero release. Stop construction of experimental fission reactors and increase spending on fusion research |
| *ca.* Feb., 1971 | Same as above | Zero release. Moratorium on construction of new nuclear power plants |
| Mar. 3, 1971 | Number of cancer plus leukemia cases may be closer to 100,000 | Not mentioned |

Sources: These data are taken from 21 position papers by Gofman and Tamplin with "GT" reference numbers; also Gofman, J., et al., 1971; Gofman and Tamplin, 1971; and Tamplin and Gofman, 1970.

dose-effect curve is more congenial to nuclear power than a "linear" curve, it is not surprising that nuclear proponents were more likely than critics to believe that the "threshold curve" was the valid one. Experts may espouse a particular position where the data are ambiguous because they are used to it and have never questioned it. An expert may take one side because his friend has taken that side, or because his enemy has taken the opposite side. In any case, these differences of opinion sometimes become very bitter, as has in fact occurred in the controversies about fluoridation and nuclear power.

Experts tend to behave like other people when they engage in a controversy. Coalitions solidify and disagreements become polarized as conflict becomes more acrimonious.[5] The same processes occur in technical controversies (Robbins and Johnston, 1976). For example, one proponent of nuclear power stated:

It's hard to maintain a detached position. I find myself forced to sweeping generalizations and extreme statements. I now find myself resenting any criticism of nuclear power, without considering the merits of the criticism (Sailor, 1971:25).

Gofman's and Tamplin's process of polarization is demonstrated by the changing position which is set forth in their books and papers over three years. Their estimates of the damage to be expected from population exposure to the Federal Radiation Council guideline of 170 millirads became increasingly higher (Exhibit 6) because they refined their models and included more effects, though there was no change in the available evidence. Their recommendations for control became increasingly stringent. Some of their opponents might regard this inconstancy as an indication of the weakness of their position, but it is simply the normal process of polarization which must be expected in any intense controversy.

## Conclusion

Technical disputes over fluoridation and radiation are confusing, in part because of rhetorical devices which obscure the problem, and in part because of arguments which talk past each other because they are dealing with different problems and derive from different premises. A calm analysis of opposing views could clear this sort of verbal thicket, but there would still remain points of ambiguity upon which experts may legitimately disagree and where it cannot be said that one is "right" and the other is "wrong."

We generally assume that informed scientific advice is valuable to political policy makers. However, in the context of a controversial political issue and when the relevant technical analysis is ambiguous, then the value of scientific advice becomes questionable. A technical controversy sometimes creates confusion rather than clarity and it is possible that the dispute itself may become so divisive and widespread that scientific advice becomes more of a cost than a benefit to the policy maker and to society. The value of technical expertise depends, in large part, on whether or not these disputes can be settled by reasonable procedures.

Several procedures have been suggested or used to resolve disputes among experts about scientific and technological issues that are relevant to policy decisions. One approach has been to suppress, discredit, or ignore technical criticism. This is imprudent because such treatment is likely to embitter and aggravate the opposition. Thus, the fact that antifluoridationists were widely attacked and discredited may help explain why fluoridation lost over 60 percent of the 1,139 referenda held in local communities between 1950 and 1969 in the United States (*Fluoridation Census, 1969*, 1970). The more acrimonious the campaign, the more likely that fluoridation would be defeated in the referendum (Crain, et al., 1969).

Another procedure would try to have the disagreeing experts resolve their own differences, or at least make their differing premises explicit. In this approach, the experts would work together in a cooperative manner, perhaps aided by a formal code of professional ethics. This may be practicable in some situations, but it is unlikely to succeed in heated controversies or when the experts are themselves seeking to bring about one decision rather than another. Then, the tendency toward polarization of the discussion into extreme positions, as in the nuclear power controversy, probably renders this cooperative approach unworkable.

A third procedure would have technical criticism evaluated in the same manner as most other scientific work, i.e., through review by scientific peers. Scientific work is typically published in technical journals after having been judged acceptable by one or several (usually anonymous) referees. Presumably, the work of, say, Gofman and Tamplin could be submitted to a few referees who would then decide to accept or reject it. The difficulty here is that, if the technical criticism is derived from ambiguous premises, the referees might reject the work simply because they do not recognize the ambiguity, or if they do recognize it, they might disagree with the critic's interpretation of it. This sort of rejection is particularly likely if the critic represents a minority viewpoint. For example, Gofman and Tamplin assume that the incidence of all cancers increases with radiation, but if that is not the prevailing view in radiation biology, many referees would probably not accept that assumption. Their critique might then be rejected, not because it was "wrong," but because the referees had outvoted the critics on how to interpret the ambiguity.

The "science court" has been proposed as another approach to the resolution of factual disputes, after they have been separated from the value differences which are usually intermeshed with them (Task Force, 1976; Mazur, 1977). Controversial issues would be referred to a science court by a means not yet specified, but perhaps through a request from a legislature or a suit in a law court or a referendum.

The procedure would begin with the selection of adversaries to represent both sides. The adversaries would be asked to state the scientific facts which they consider most important to their respective cases, and to supply documentation to support their assertions. They would then exchange their lists and documentation, examining each other's claims and specifying those assertions with which they agreed and those with which they did not. A referee who was acceptable to both adversaries would attempt to arbitrate differences between them, perhaps by alterations of wording or by the removal of ambiguities in factual statements. In the event that both adversaries agreed on all listed statements of fact, the procedure would end, and these facts would constitute the science court's report.

If one or both sides challenged factual statements by the other, these challenges would be the subjects of a hearing, open to the public and governed by a disinterested referee, in which the adversaries argued their opposing scientific positions before a panel of scientist-judges. The referee would restrict the debate to questions of fact, excluding issues of moral judgment and policy. The judges themselves would be established experts in fields which were relevant to the dispute, but would not be drawn from scientists actually working in the area of dispute, nor would they include anyone with a personal bias or organizational affiliation which would predispose him towards one side or the other. The judges would be subject to challenges by the adversaries on the grounds of prejudice, and the court would only proceed once a panel of judges had been selected which was acceptable to both sides.

After the scientific evidence had been presented, questioned, and defended, the judges would prepare a report on the dispute, noting points on which the adversaries agreed and reaching judgments on factual issues still in dispute. The judges might be able to decide if either or both of the adversaries were wrong, or if the differences between them were legitimate, resting on points of irreducible ambiguity or insufficient data. They might suggest specific new research to clarify points which remain unsettled.

The judges would not make a recommendation on policy, or a judgment of moral right or wrong, since they are assumed to have no special wisdom or political mandate that permit them to decide moral issues in their society. They should, however, be particularly qualified to arrive at purely scientific conclusions, and this would be the limit of their mandate. The court's report could then serve as the factual basis for value judgments by other bodies more properly charged with policy-making functions, such as the legislature or a regulatory agency or the public in the case of a referendum.

This adversary procedure has flaws too. A particularly persuasive adversary might sway the judges more by his oratory than by his evidence, just as a successful trial lawyer can win a jury to his side more by appeals to sympathy than to logic. For example, John Gofman is an excellent advocate and on that basis alone he might be more successful before a panel of judges than would be a less eloquent speaker.

One particular appeal of the science court is that factual disagreements would be separated clearly from disagreements over value questions. Factual disputes may sometimes be resolved by the analysis of data and so can be treated differently from value disputes which are impervious to the methods of science. Other procedures, aside from the science court, have been proposed to treat fact and value disputes separately. Casper (1976) has suggested that the public would be better informed about technical controversies if the television channels carried separate debates over the factual and value disagreements in such controversies, rather than intermingling them, as they do now. Elite groups, such as professional associations, or seminars of scholars from the humanities as well as science and engineering, might improve their insights into these controversies by examining the value disagreements apart from the factual disagreements. It is worth examining the feasibility of this approach, which is the task of the next chapter.

## Notes

1. This research is reviewed in Crain, et al. (1969).

2. A counterargument, claimed by Gofman and Tamplin, is that three percent of the cases of cancer and leukemia are caused by naturally occurring background radiation (GT-102-69, 1969:12-13). "GT" reference numbers refer

to a mimeographed set of position papers by Gofman and Tamplin, available from these authors.

3. A 1952 congressional committee, one of the only informed bodies to take the low-dose fluorine "danger" seriously, had been investigating the dangers of chemical food additives for a year before it took up the question of fluoridation. It is not surprising that the committee was extremely sensitive to the possible toxic effect of adding a chemical to the water (*Hearings*, 1952).

4. All of the mice were normally tumor prone. The mice receiving fluoridated water developed tumors sooner than the control mice, but their total incidence of tumors was no higher than in the control group. Confusion on this point evidently led to the belief that fluoridated water causes cancer. Taylor explicitly denied that his data showed any increase in cancer incidence (*Hearings*, 1952:1540).

5. See Coleman (1957), Coser (1956), and Mazur (1968) for discussions of polarization.

# 3
# Separating Disputes over Facts from Disputes over Values

*(with Andrew Marino and Robert Becker)*

The classical argument over the relationship between facts and values has been extended to technical controversies, with critics objecting that a clear separation is impossible. This argument has obscured more than it has clarified, since it suggests that a total separation of statements of fact from any evaluative statement is necessary. That is not true. All that is required is a separation of blatant evaluative or normative statements from statements of fact. Values which are shared by all the contending interest groups, or values which are too subtle to affect practical decisions, may be intertwined in the statements of fact without causing a problem.

A practical decision to build a nuclear power plant can easily be analyzed into factual questions, such as "How many cancers-per-year will be produced in the population exposed to radiation?" and into normative questions, such as "How many cancers-per-year should be accepted in exchange for the amount of electricity generated by this plant?" The problem is not so much that questions of fact and value are not empirically separable, but rather that an adversary may find it rhetorically useful to state his factual hypotheses in terms which make them difficult to evaluate.

It is intrinsically difficult to demonstrate the absence of an effect or the impossibility of an event. An opponent of apple juice might argue that apple juice, consumed at the rate of five gallons per year per

---

This chapter is adapted from an article by the author with Andrew Marino and Robert Becker, entitled "Separating Factual Disputes from Value Disputes in Controversies over Technology," which originally appeared in *Technology in Society*, Vol. 1, 1979, pp. 229-37. Used with permission.

person, has harmful effects on the human organism. Proponents of apple juice would probably dispute this statement, which might then be argued before the judges of a science court. Consider now the positions of the judges: in order to reject the hypothesis, they must have evidence of "No effect" on all conceivable forms of mortality and morbidity. This is nearly impossible. Furthermore, even for the forms of mortality and morbidity which the defenders of apple juice did examine, the most that they would be able to say would be that they *discovered* no harmful effects, which is a weaker statement than that there *are* no harmful effects. The problem here is the intrinsic difficulty of proving the nonexistence of something—even if it does not, in fact, exist.

The report of a science court would certainly not affirm the dangers of apple juice, but neither could it reject the hypothesis outright. The report might, therefore, be interpreted as casting doubt on the safety of apple juice, even though there is no basis for such concern.

Adversaries might find it useful to state their hypotheses in vaguely probabilistic terms which contain no clear-cut criteria for realistic assessment. It might, for example, be asserted that "ionizing radiation probably increases all forms of cancer in humans." "Probably" is a vague term; judges have no clear criterion for deciding whether or not an effect is "probable." More precise criteria are easier to evaluate, but may sometimes be as misleading as vague terms.

Suppose the hypothesis had been stated as follows: "A population exposed to 170 rads/year of ionizing radiation will have significantly more colon cancers than a similar population not exposed." "Significant" has a precise meaning in statistics, i.e., that the probability of erroneous rejection of the "null hypothesis" is less than .05. If $p =$ .06, then the null hypothesis is—in the technical sense—sustained, even though this result would clearly suggest that radiation actually did cause the cancers.

We must expect these kinds of rhetorical devices to appear in the factual statements which adversaries submit. For example, a scientist-opponent of the supersonic airplane has suggested that a science court examine this hypothesis: "In 1971 there was sufficient scientific evidence to establish *probable cause* . . . that 1.8 million tons per year of nitrogen oxides [as $NO_2$] injected by supersonic transports . . . at an elevation of 20 kilometers would reduce stratospheric ozone by . . . a global average of up to 20 percent . . . ." At first glance, this looks like a precise, factual statement. But the phrase "probable cause" is vague, with no clear criterion for deciding whether or not it could have been established.[1] Furthermore, the claim that ozone is reduced "*up to 20 percent*" is necessarily sustained with any reduction greater than 0 percent, no matter how small. In order to refute this claim, one would have to demonstrate the nonexistence of any reduction. Thus, the hypothesis has been stated in a way that there is little risk of the claimant being shown to be wrong.

It would be a great error to assume that adversaries purposively distort facts as a ruse to support their own positions, though it would be naive to believe that this *never* occurs. In particular, it is wise to take into consideration the pressure on a technical expert who is engaged in a heated controversy, especially if he is opposing an "establishment" position with few resources of his own. There is a common tendency in such a situation to take a defensive posture, stating one's technical position in a manner that provides little opportunity for a clear refutation by the other side.

When such statements appear on both sides, as is frequently the case, the task of sorting out *who* is saying *what* is exceedingly difficult for the layman. The sophisticated judges of a science court would have an easier time sorting out the claims, but the difficulty of evaluating such statements is great. Obviously, the substance of scientific dispute must be stated in a way which allows meaningful assessment through scientific methods. This is true of any other procedure which requires the separation of factual disputes from value disputes.

## The Role of the Referee

In the science court procedure, a referee has the task of obtaining factual statements with supporting documentation from the adversaries, determining which statements are accepted by both sides and which are not, and attempting to mediate differences between the sides. No doubt, much of the mediation will focus on problems of wording, such as are discussed above, so that the substance of the factual claims will be stated in a manner that allows fair assessment, allowing either a confirmation or a refutation by reasonable scientific means. It would appear that procedures other than the science court would have the same problem in separating factual disagreements from value disagreements, and, therefore, would require some mechanism to function as a referee.

Given the central importance of the referee function, it is disturbing that we have little sense of the feasibility of performing this role. The relationship between adversaries is usually hostile, and the referee has no sanctioning power with which to enforce the cooperation necessary for the production of workable factual statements. It would not be surprising, under these circumstances, to find that each side's statements are so "loaded"—so prohibitive to adequate testing—that the entire procedure falls apart.

In order to assess the workability of the referee's role, I attempted to promote an exchange between technical experts who were involved in a controversy over high-voltage transmission lines.

## The Transmission Line Controversy

Citizens' groups in upstate New York and in Minnesota have opposed the construction of new, very high-voltage transmission lines. As in most other technical controversies, objections include both technical and nontechnical issues. Opponents of the proposed construction are afraid that the lines are more dangerous than the utility companies realize, or will acknowledge, and there has been dispute over the magnitude of the hazards associated with extremely low-frequency electromagnetic fields of the type which these installations would produce. Technical presentations by both sides of the controversy have appeared in the permit hearings of the New York State Public Service Commission. More- or less-adequate summary accounts have been reported in the mass media.

Quite apart from this scientific issue, there have been objections to the use of the power of eminent domain by the utility companies to enforce the purchase of rights-of-way from farmers who are reluctant to sell sections of their land; this is seen as infringement of individual property rights. Other issues which have been raised are that (1) construction of the lines would promote the related construction of nuclear power plants; (2) utility companies do not serve the public interest; and (3) certain elected officials have not properly performed the duties of their offices. In Minnesota, particularly, there have been destruction of property and incidents of physical violence between the protestors and national guardsmen who were sent to protect the construction sites.

Drs. Andrew Marino and Robert Becker, of the Veteran's Administration Hospital in Syracuse, New York, have been technical critics of the lines, and have argued their case in public hearings and other community settings. I attempted to act as a referee between Becker and Marino, the salient technical opponents, and four experts who were closely associated in hearings and published accounts with the argument that fields from the transmission lines are not hazardous.[2]

## The Procedure of the Referee

Could a referee elicit from experts testable statements of alleged facts which seemed relevant to the policy issue, but were relatively divorced from strong value implications? Acting as referee, I wrote to each expert involved, identifying myself as a nonpartisan, and explaining my interest in the transmission line controversy as a test case for a science court; I enclosed an article about the science court from *Science* magazine. I asked for aid in constructing and critiquing a list of alleged statements of fact which are under dispute.

Most of the experts contacted did respond, although some of them required reminders before they did so. Most of them, apparently, read the *Science* article, and made some effort to construct or criticize the facts lists. However, only opponents of the lines produced explicit lists. The other experts cooperated mainly by providing me with critiques of those lists, which I then channeled to Becker and Marino. Becker and Marino then prepared a revised list in response to the criticisms, which I sent to the pro-line experts along with a request for comments. Three of the four experts who received the list responded.

Of course, this procedure was only a rough approximation of that which would occur in the initial phases of an actual science court. But it did carry out the crucial attempts to enlist the cooperation of experts who—in some instances—regarded one another with enmity, and also to obtain fairly worded statements which would set out the areas of factual disagreement. What, then, was the result of this attempt?

## Revision of the Lists

The strategy of Marino and Becker in compiling their original list was to assert first that biological effects from transmission line fields were *possible,* since they believed that at least one of the pro-line experts had maintained that such effects were not possible. This done, they then asserted that biological effects were not only possible, but *likely.* Finally, they argued that such likely effects cannot be shown to be safe (and therefore, it is inferred, may be dangerous).

Here are the explicit points of their argument:

1. Extremely low frequency (ELF) electric (and magnetic) fields *can* cause biological effects in human beings exposed thereto.

2. It is likely that ELF electric (and magnetic) fields associated with high-voltage transmission lines will cause biological effects in human beings exposed thereto.
3. No biological effect that is likely to occur in human beings exposed to the fields of high-voltage transmission lines can be shown to be nonhazardous.

This list was criticized by most of the pro-line experts as vague and untestable. The first statement, asserting that the fields *can* cause *biological effects* in humans, was criticized because they felt that the term "biological effects" is too broad to be meaningful in an empirical sense, given the wide variety of such effects which would have to be examined. Also, the statement that the fields *can* have effects is irrefutable, because opponents would have to show the converse: that fields *cannot* have effects. This is the problem of nonexistence which was discussed earlier. This same problem appears in the third statement which requires for refutation a proof of the nonexistence of hazard. The second statement maintains that field effects are *likely,* but there is no clear criterion for assessing whether or not an effect is "likely."

This initial list of statements shows the tendency—a common one in acrimonious dispute—to make ostensibly substantive empirical claims which are logically or pragmatically irrefutable.

To appreciate the impact of the exchange of statements and critiques among the experts, one must be aware of the fact that the opposing sides had had virtually no direct contact during their long involvement in the controversy (from 1974 until 1977), and they never before had been called upon to compare their scientific positions on a point-by-point basis. An important misconception soon became apparent. Becker and Marino had attributed to one pro-line expert the view that it was impossible to produce biological effects from low-intensity fields—a view that that expert denied to me, acting as referee, that he held. Thus, at least one point of dispute was settled by the exchange.

Becker and Marino prepared a revised list in an attempt to respond to the critiques of the first list; this list appears in Exhibit 7. Here their allegations are phrased in the form of epidemiological hypotheses with a degree of specificity that is common in standard journals in that area. They alleged that people exposed to fields created by transmission lines (of a given design) for a period "as short as five years" will differ from a control population not so exposed in several enumerated characteristics.

EXHIBIT 7

Revised List of Facts Relevant to Siting Electrical Transmission Power Lines

People exposed for a period as short as five years to the electromagnetic field created by a 765 kV transmission line (as specified, for example, in "Application to the State of New York Public Service Commission for Certificate of Environmental Compatibility and Public Need," submitted by Rochester Light & Electric Corporation and Niagara Mohawk Power Corporation, January 1974) will be more likely to differ from a control population not so exposed in the following characteristics:

1. Growth, as measured by rates of change of physical parameters (e.g., height, weight).
2. Biological stress, as measured by physiological indicators (e.g., corticoids, serum proteins, circulating lymphocytes, blood pressure) and incidents of stress-related diseases (e.g., gastrointestinal and cardiovascular disorders).
3. Functioning of the central nervous and cardiovascular systems, as measured by neurohormone patterns, EEG, EKG, and the ability to adapt to blood volume changes.
4. Psychological behavior, as measured by decision-making capability, rates of acquisition of learned responses, gross activity level, reaction time, short-term memory, and motor coordination.

A refutation of this type of statement no longer requires the impossible demonstration that these fields *cannot* cause effects. An adequate refutation would consist simply of a comparison of exposed and unexposed populations and a demonstration that differences between them are no greater than one would expect by chance, using conventional levels of statistical significance. (The statement could be improved by specifying the details of such a study, including significance levels which would be accepted as test criteria).

Becker and Marino attempted to add specificity to their statements by listing four broad biological variables as effects and then giving specific indicators which are conventionally measured in physiological and psychological research on these broad variables. One such broad variable, and its exemplar indicators, is "Growth, as measured by rates of change of physical parameters (e.g., height, weight)."

One pro-line expert wrote that the revised list was still "almost impossible to judge . . . on the basis of current scientific knowledge or through a reasonable, empirical experiment." He suggested cogent improvements, such as a better specification of both the level of the electromagnetic field and the period of exposure alleged to cause effects. He also pointed out that the specification of exemplar indicators for the broad biological variables leaves uncertainty because, if no differences appeared on these exemplars, Becker and Marino could respond that additional indicators should be examined, and they would again be in the bind of having an infinite variety of indicators that might be considered. This expert suggested a complete specification of the indicators that would be examined; these modifications could be incorporated into a further revision.

Two other pro-line experts found the revised statement sufficiently specific that they could disagree that the allegations were true for humans. The fourth pro-line did not comment on the revised list.

The statement of alleged facts could be improved further, but there is little doubt that the revised list is a substantial improvement over the initial list. These results support the contention that a referee can obtain from opposing (and hostile) experts a list of alleged facts which are empirically meaningful, which are points of disagreement, and which are reasonably separated from the policy decision. In this case, the appraisal of biological effects from electromagnetic fields can proceed independent of decisions on whether or not to build transmission lines, or on levels of risk which are "acceptable" to the public.

## Politics Dominate Scientific Inquiry

There is one more lesson in this largely-successful attempt to separate the factual disputes of the transmission line controversy from its value disputes. Once done, it was not possible to bring the experts together to debate their positions on factual issues; the pro-line experts as a group did not want any sort of involvement in a science court procedure.

Why were the pro-line experts reluctant to act as adversaries? One of them objected to the idea of a science court: "The concept of a 'Science Court' is foreign to the scientific method, . . ." "the adversary approach . . . is anti-science." One of them objected specifically to Marino as an opponent, claiming that "his arguments and conclusions are sophistic."

Some of the pro-line experts thought that the public exposure of a science court was more likely to hurt than to help their side. One said that it "may impute scientific credence to Dr. Marino's arguments which they do not deserve." Another said that the transmission line issue should not be considered in an untried science court, because it might "go wrong." One wrote: "It would be unfortunate to find that, as a result of untested procedures . . . the science court might directly contradict the deliberate considerations of these same issues by well-established bodies such as the National Academy of Sciences . . . The potential impact upon the Nation's system of transporting electrical energy which might result from a Court decision inconsistent with those of other government agencies ought to be seriously considered in determining whether these issues are ripe for science court arbitration, particularly in the absence of any prior experience with such a court."

In the transmission line dispute, it is the proponents of technological development who object to a debate of scientific disagreements, in part, because it publicizes and perhaps legitimizes environmental criticism of the lines. Why, after all, enter the debate if it is more likely to improve the relative position of the other side than of one's own side?

This response emphasizes that many technical controversies are primarily disputes over political goals and only secondarily concerned with the veracity of scientific issues which are related to these goals. Why, then, should anyone care about resolving factual disputes? If the final report of a science court would probably not alter the positions of the adversaries and their interest groups, why bother?

One reason is that a reasonably sophisticated and relatively unbiased report on the factual matters in dispute could have an important impact on that portion of the public which has not yet taken a side in the controversy, but whose interests are at stake. If, in the controversy over the fluoridation of drinking water, a science court reported that there were, indeed, grounds to doubt the safety of fluoridation, then it is likely that previously uninvolved people would join the opposition to fluoridation in their communities. If the technical objections raised against transmission lines or nuclear power plants were found to lack any scientific basis, and this was reported by a credible source, then political power would most likely shift to the proponents of these technologies as electricity became scarcer and more expensive, and previously nonaligned citizens became involved. The resolution of factual disputes may not serve the interests of those directly involved in the debate, but it would be in the best interests of the public at large.

## Notes

1. Johnston (1976). A footnote adds, "In legal matters, *probable cause* is sufficient for a grand jury to recommend that a case be tried in a court of law, rather than be dismissed.

2. During this exercise, I acted independently of Marino and Becker. Afterward, we three decided to share the lessons learned during the dispute.

# 4
# Partisans

The last two chapters focused on the scientific arguments which are found in technical controversies. These next three chapters will look at the people—scientists and nonscientists—who become involved in the controversies. Why do some of us favor nuclear power plants while others, equally intelligent and concerned with the public welfare, oppose them? How does anyone, expert or layman, choose a side in a controversy? Who are the people who become actively involved?

My purpose in this chapter is to review available data on partisans in order to suggest generalizations which may be true across most technical controversies. Before I attempt to describe the partisans, it is necessary to differentiate between active participants in the controversy and passive members of the wider public who simply express their views in a referendum or in response to a question on an opinion poll, or who spend a day listening to rock music at a demonstration. The demographic and motivational characteristics of these two classes of partisans may be quite different, a point which is lost in some of the early literature of this field. Activists are a very small portion of the total population, so regular national sample surveys cannot be used to infer their characteristics. Specific studies of activists are needed, and unfortunately, few exist.

Almost invariably, two distinct sides emerge in any technical controversy which has escalated to the point of widespread public attention, and though one can often find more than two factions, these factions will be indentifiable with one side or the other. It is usually convenient and accurate to label these the "establishment" position, and the "challenge" position which opposes the establishment. The establishment position is associated with some combination of government offices, perhaps the military, corporate industry, and professional organizations; while the challenge side is associated with voluntary organizations such as environmental and consumer groups, religious organizations, or ad hoc groups formed specifically to promote this protest or a set of related protests. The challenge side may include some establishment figures, such as influential congressmen, but in these cases it is clear that these people do not represent an establishmentarian point of view.

The establishment side usually supports the technology while the challenge side opposes it, either wholly or in its proposed implementation. Thus, the establishment position supports nuclear power plants, fluoridation, the supersonic transport, the ABM, high-voltage electric transmission lines, and recombinant DNA research, while challenge groups oppose them. There are exceptions, as in the Laetrile controversy, where the establishment (particularly the medical profession and the Food and Drug Administration) opposes the legalization and use of the drug, while challenge groups support it.

## Occupational Activists

Most activists on the establishment side of a controversy have become involved through the occupational activity of their workday jobs. Employees of electric utility companies favor nuclear power, physicians oppose Laetrile, dentists favor fluoridation, aerospace engineers supported the supersonic transport. Exceptions to this pattern are so rare as to be newsworthy, as when four nuclear engineers left their jobs to join the antinuclear protest movement (Gwynne, et al., 1976). From the perspective of most employees of the corporate or governmental institutions which supports some technology, those institutions behave in a responsible and trustworthy manner and their advocacy of a technology is compatible with—even beneficial to—the public interest; those who oppose the technology may be seen as misguided citizens who do not understand, or perhaps simply as troublemakers. A number of factors encourage this perspective. The establishment employee earns his livelihood from the support of the technology (directly or indirectly), his career may be tied to the viability of the technology or at least to the businesses and agencies that support it, and his work day is spent among friends and associates who are like minded. The common pattern in this situation is to support the policy of one's organization, unless perhaps one is having a particularly bad time on the job, in which case joining the protest may be a convenient outlet.

Though most employees may favor the technology, only a few become active in the controversy, and typically this activism is part of their jobs. As examples, proponent activists are likely to be the public relations officer or legal counsel or a senior executive of an involved corporation, or the president of a trade or professional association, or the chairman of a government agency. No doubt additional factors come into play. Of several prominent officers in the nuclear power industry, a few are particularly articulate and persuasive under the pressure of confrontation, and once gaining some experience in debate and some support from their constituency, these people often emerge as major spokesmen in the controversy.

The demands of one's occupational role also account for some activism on the challenge side of the controversy, for among the voluntary groups is a minority of paid professionals including writers, organizers, and officers of environmental and consumer organizations, and lawyers specializing in regulatory and consumer affairs. These professionals are essentially carrying out the demands of their careers in the course of protest, just as the establishment professionals pursue their careers via advocacy of the technology. In this sense, the behavior of professional challengers is more similar to their establishment adversaries than it is to their allies who protest as a voluntary sparetime activity. We can visualize two bright high-school students, one choosing a career in engineering and on graduation accepting a job in the electric power industry, the other choosing a career in law and on graduation joining a firm which specializes in consumer affairs. Ten years out of high school, they may face each other across the table as adversaries in a nuclear power plant licensing hearing. Why one person chooses engineering while another chooses law is not well understood, though social background variables beyond their own control probably had some influence in career selection. Prior ideological commitments may have led the lawyer into consumer affairs rather than corporate law, but luck and limited job opportunities also channel people to one particular employer rather than another. Once embarked on their chosen career lines, the professional and organizational influences which led the lawyer and the engineer to their respective sides of the hearing table are not difficult to comprehend.

## Volunteer Activists

While the vast majority of activists on the establishment side have become involved in the controversy and participate in it through the demands of their occupational roles, this is not true of the challenge side where with the exception of the few paid professionals mentioned earlier, activists contribute their time and other resources outside of their regular work activity, for little or no pay.

There is little good data on the active opponents of various technologies. Much of the material that is available has been produced by the people on the other side and is not trustworthy. For example, proponents of fluoridation have fostered the view of their opponents as cranks, bigots, and radical right-wingers in quixotic battle against an imagined communist conspiracy (American Dental Association, 1965). This may be true of a few individuals but is not accurate as a general characterization of antifluoridationists.[1]

I have taken the most reliable available data on antifluoridation activists,[2] and my own interviews of antinuclear activists,[3] to form the following comparative picture.

Both groups tend to be of middle age or older, they are of both sexes, they are relatively well educated, and they were often active in other public affairs before becoming involved in their technical controversy. Many have occupations which allow them to devote considerable time to the controversy, for example, retired men or housewives with grown children; or writers, teachers, or scientists in fields related to the technology. They are knowledgable about the technology which they oppose, and they consider its hazards to outweigh its benefits. They express their opposition to the technology in the context of larger ideological concerns: antifluoridationists fear the power of central government to impose mass medication; antinuclear activists are concerned about the degradation of the environment and corporate control of society. In most of these regards and others, they are similar to the scientist-activists in the ABM controversy (Cahn, 1974; though most of these were men), to active antiabortionists (Leahy, 1975; Mazur and Leahy, 1978), and also to the active volunteer spokesmen for most other American political issues (Verba and Nie, 1972).

The activists of each movement have a distinct political cast, though few are at the extreme ends of the political spectrum, as is sometimes claimed. Antifluoridationists are usually conservative. Antinuclear activists are usually liberal. Anti-ABM scientists also tend to be liberal, as indicated by their peace posture on the Vietnam War and their antipathy toward President Nixon at the time of the controversy.

The politics of each movement's activists needs to be explained. None of the technologies being opposed is intrinsically liberal or conservative, nor are many of the ideological issues that surround them. Liberals oppose big business and big government in the controversies over nuclear power and the ABM, and conservatives take a similar position in the controversy over fluoridation. The environmental issue, so prominent in the nuclear power debate, has appealed to conservatives as well as liberals. Why, then, are the activists of each movement politically partisan?

The political complexion of each movement was set early in the controversy and remained fairly stable throughout its course. The left orientation of the opponents of nuclear power plants could be observed as early as 1956 when the United Auto Workers of America and two other unions intervened against the Enrico Fermi nuclear plant near Detroit. This appears to have been, in part, a political attack on the Eisenhower administration which was promoting nuclear power reactors in its Atoms for Peace program. Once these initial lines were drawn, liberals were more likely than conservatives to find the antinuclear issue appealing, and they were more likely to be recruited by the early liberal activists. Thus the social process of recruitment maintains the political complexion of the antinuclear movement's early leaders.

As late as 1967, President Johnson and Secretary of Defense Robert McNamara, who had become targets of liberals because of their Vietnam policy, opposed deployment of an ABM. Johnson and McNamara suddenly reversed their decision in September, 1967. This provided the antiwar liberals in Congress and the universities with a convenient new issue on which to challenge the Johnson administration, and opposition to the ABM became tied to opposition to the War.

I have not been able to reconstruct the first fluoridation disputes in sufficient detail to explain why the early proponents were conservatives. One hypothesis is that the small but vociferous group of Wisconsin dentists who pushed fluoridation were, by chance, liberals who would have been opposed by community conservatives no matter what the merits of their proposal (McNeil, 1957).

For lack of data, I have not been able to draw a very complete picture of the active partisans, but it is sufficient for us to make comparisons with partisanship—to the slight degree that it is expressed—within the general public.

## The General Public

National opinion polls almost always show support for the estab-

EXHIBIT 8

Percentage of Public with Great Confidence in the People
Running each of Nine Institutions

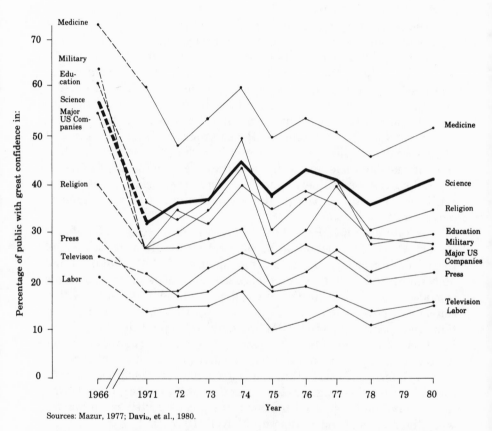

Sources: Mazur, 1977; Davis, et al., 1980.

lishment position in a technical controversy. Polls taken throughout
the 1950's and 1960's showed the opinionated public (i.e., those who
express an opinion) in favor of fluoridation by four-to-one or more.
During the ABM controversy, a majority of the opinionated public
consistently favored deployment. National opinion polls continually
showed majority support for nuclear power plants until the accident
at the Three Mile Island nuclear plant in early 1979. A few of the
many polls taken since the accident showed the public nearly evenly
split on nuclear power, but most showed that technology maintaining

a majority of the opinionated public, particularly as the accident receded in time.[4] One can bring these support percentages down by manipulating question wording,[5] but it is clear that the American public tends to favor the establishment position in a technical controversy.

Widespread popular support for science and technology has been documented repeatedly, contrary to continual erroneous claims that they have fallen into disrepute (Marshall, 1979). In 1966 and then in 1971 through 1980, random samples of the population were asked how much confidence they had in the people running various institutions in America, including science (Exhibit 8). Confidence in all institutions was higher in 1966 than during the 1970's, though we have no idea if 1966 was a "typical" year or one when confidence was unusually high. In any case, through the 1970's science has done well in the public mind, with about 40 percent of respondents reporting "great confidence" in its leaders, and only about 7 percent reporting "hardly any confidence" (Davis, et al., 1980). Science has shown little

EXHIBIT 9

Correlations (Qs) between Attitudes toward Fluoridation or Nuclear Power Plants and Demographic Variables

| | Fluoridation (opposition scored positive). | | | | |
|---|---|---|---|---|---|
| Year | 1952 | 1953 | 1956 | 1965 | 1968 |
| n | (1222) | (988) | (1196) | (1234) | (1064) |
| Sex | | | | | |
| (male scored +) | +.08 | −.03 | +.10 | +.04 | +.29* |
| Age | +.39 | +.38 | +.47 | +.40 | +.15* |
| Education | −.16 | −.14 | −.11 | −.15 | −.33* |

| | Nuclear Power Plants (opposition scored positive) | | | | | |
|---|---|---|---|---|---|---|
| Year | 1956 | 1971 | 1972 | 1975 | 1976 | 1979 |
| n | (661) | (ca. 1200) | (1326) | (1537) | (1497) | (1200) |
| Sex | | | | | | |
| (male scored +) | −.34 | −.32 | −.33 | −.33 | −.23 | −.33 |
| Age | +.03 | +.01 | +.05 | −.11 | +.07 | +.04 |
| Education | −.31* | −.12 | −.07 | +.04 | −.09 | −.13 |

Sources: Fluoridation pools were obtained from The Roper Public Opinion Research Center, Williamstown, Mass. (control numbers AIPO 0484K0, AIPO 0514, and AIPO 0559) and National Opinion Research Center, Chicago (survey numbers SRS-868 and SRS-4050). The nuclear power poll for 1956 was obtained from the Roper Center (control number AIPO 0558), the polls for 1971 and 1972 were obtained from the Electric Companies Public Information Program, New York, and the polls for 1975 and 1976 are reported in Harris and Associates (1976), the poll for 1979 (taken April 6-9, about one week after the Three Mile Island accident) is from ABC News-Harris Survey.

* Anomolous values (see text).

loss throughout the 1970's in the percentage of respondents who place great confidence in its leaders, and its ranking relative to other institutions has improved.

However, the public's bias in favor of establishment science and technology does not mean that its support is inevitable. If protest is sufficiently vocal and focused, the public will move to opposition. This shift to the challenge position is less apparent in opinion polls of the total population, which include its inert apathetic elements, than in referenda where the most concerned elements of the population are over-represented among the voting minority of the country (Hensler and Hensler, 1979). Fluoridation lost over 60 percent of 1,139 referenda held in American communities between 1950 and 1969, and defeats were particularly likely when campaigns were acrimonious, bringing the protest forcefully to public attention (Crain, et al., 1969).

Time series data are sparse on public reactions to controversial technologies. However we may get some notion of the structure of public opinion from a potpourri of polls which were taken during the long running controversies over nuclear power and fluoridation. Several for which breakdowns were available are listed in Exhibit 9. These polls vary in questionnaire wording, polling organization and sample frame, though all are large national-level surveys allowing breakdowns of opinion by the demographic variables: sex, age, and education. To ease comparison across the polls, the strength of the relationships between each demographic variable and opinion toward the technology is measured by the correlation coefficient $Q$ which takes values from $-1.0$ to $+1.0$. Values near zero indicate that the variables are independent, while higher values of $Q$ (either positive or negative) indicate that one variable may be predicted from the other. Positive values of $Q$ would indicate that opposition is highest among men, older respondents or better educated respondents; negative values indicate the reverse.[6]

The specific correlations for a given controversy in a given year are of little interest, but they do suggest generalizations which may be broadly true across all technical controversies. The first pattern to be noted is that for each controversy, the correlations for each demographic variable are stable from one poll to the next. There are a few values of $Q$ which depart from this pattern of stability. These are asterisked in Exhibit 9, where most are due to the 1968 fluoridation survey which is anomalous. Thus, sex is almost always a poor predictor of attitude toward fluoridation (i.e., $Q\approx0$) except in 1968, when men were somewhat more opposed than women. However, women invariably show greater opposition than men to nuclear power (Reed and Wilkes, 1980). Older respondents are always less favorable to fluoridation than are younger respondents, but age never predicts nuclear attitude to any substantial degree. Poorly educated respondents are slightly more opposed to both technologies than are well educated respondents, in almost every poll. The robustness of these relationships is remarkable given the heterogeneous nature of these polls, and the long time spans between them. The correlates of opinion which are set early in the controversy seem to persist throughout.

A second generalization of interest is that the pattern of demographic correlates is unique for each controversy, a point which becomes clearer if we extend the comparison to the short-term ABM debate. Women are particularly opposed to nuclear power plants but not to fluoridation or the ABM. Older respondents are particularly opposed to fluoridation but not to nuclear power or the ABM. The ABM controversy is the only one of the three in which opposition was greater among well-educated respondents than poorly educated respondents (though not by much). Thus the public cannot be regarded an amorphous mass society which gives the same undifferentiated response to every technological innovation which is proposed. To the contrary, the public's response to each technology is highly specific. It seems likely that this is a reaction to particular features of the technology or the associations it evokes rather than to some generalized image of technological change.

## Differences between Activists and the General Public

The people who are active on either side of a technical controversy are a very small portion of the total population, and a nonrepresentative portion. They are from the higher rather than lower strata of society, usually middle age or older, and often active in community affairs. In contrast, most members of the general public are not very concerned about technical controversies (or other public issues), and those who hold an opinion express it quietly, usually in support of the establishment position, though a dramatic event which is widely reported in the mass media (e.g., Three Mile Island) can greatly increase public attention and opposition for awhile. Most of the time, however, the action of a technical controversy is played out by the activists on either side, who are few in number.

Failure to distinguish between activists and the general public has led to analytical errors. For example, some proponents of nuclear power and fluoridation, as well as some social scientists, attribute the opposition which they encounter to ignorance, as if people would favor the technology if only they understood it (Metzner, 1957; Doff, 1970). This belief seems to be based partly on the opinion polls which show that less educated people hold relatively negative views about both fluoridation and nuclear power. An inference is made that uneducated people do not understand the technology, and that is why they oppose it.[7] The explanation is stretched further by assuming that what may be true of the general public is true of activists as well, i.e., that activists who oppose the technology do so because they do not understand it (Seeley, et al., 1971). This reasoning fails because it confuses activists with passive members of the general public, attributing similar characteristics to these two very different groups. It is quite clear that most active opponents of a technology, having spent a great deal of their time in study and debate, are quite knowledgable about it, certainly more so than the general public.

Another difference between activists and the general public, one that will be important in the subsequent discussion, is that activists embed their positions for or against a technology in a larger ideological framework of social and political beliefs, while most members of the general public do not. All studies of antifluoridation activists show that their opposition has been grounded in an articulated, general outlook, usually embracing individual freedom of choice, opposition to socialism, and the notion that powerful central government is swallowing up the rights of the individual (McNeil, 1957; Davis, 1956; Green, 1961; Mueller, 1966; Gamson, 1968; Crain, et al., 1969). The profluoridation activist speaks of spreading health benefits to all mankind, the poor as well as the rich. Antinuclear activists explain their concerns as an expression of environmental beliefs, particularly regarding pollution, ecology, energy, and corporate control of society (Mazur, 1975; Holden, 1979). The pronuclear activist is concerned about maintaining the industrial and economic growth that have made Americans prosperous and the nation great.

As important as these beliefs about larger issues are to activists, they have little relevance to most of the general public. For example, several studies of fluoridation, using attitude surveys and referendum data, show little connection between political conservatism and opposition. My own attitude survey of Syracuse, New York, showed no relationship between attitude toward nuclear power plants and concern with population problems or environmental issues (Mazur, 1975).

As a corollary of this ideological difference, activists talk about a technology, and explain their posture toward it, very differently than do most respondents from the general population. The activist has thought at length about the technology, its risks and benefits, and how his support or opposition to it fits into his larger world view. It is an issue which he has discussed in detail, usually with adversaries as well as allies, so that his own position is well articulated and highly defensible, much more so than the position of a passive partisan. In the following chapter we will look at the process whereby an activist's position for or against a technology may become meshed with his ideological rationale for that position.

## Notes

1. Mueller (1966) encountered the idea of a "communist conspiracy" in only one of seven fluoridation campaigns he studied, and Groth (1973) did not find it in any of the five which he studied. Several prominent scientists are included among the antifluoridationists (Groth, 1973; Jerard, 1968).

2. Two statistical descriptions of antifluoridationists are helpful here, but both reflect severe biases. Crain, et al. (1969) interviewed local newspaper publishers to obtain information on leaders of community fluoridation challenges. Since these publishers were predominantly profluoridationist, their descriptions almost certainly overstate protestors' negative points: 33 percent called the antifluoridationists "cranks." *The National Fluoridation News,* an antifluoridation newspaper, conducted a poll of its readers in 1971 and received returns from 570 respondents who probably form a disproportionately well-educated sample of people persistently opposed to fluoridation, many of them presumably activists (Groth, undated). A discussion of the reasonable inferences that may be drawn from these two studies appears in Mazur (1975).

3. In early 1973, I interviewed 30 leaders of organizations belonging to National Intervenors, a coalition of groups opposed to nuclear power plants. Details appear in Mazur (1975).

4. References for most of the nuclear power and fluoridation polls discussed in this section appear in Exhibits 9 and 21. ABM public opinion was measured in three Gallup polls available from The Roper Public Opinion Research Center, Yale University. These were taken in March (AIPO 777), May (AIPO 780), and July (AIPO 784) of 1969. For question wording see Chapter 8, footnotes 1, 2, and 3. See Farhar, et al., 1977, and Melber, et al., 1977 for reviews of opinion research on nuclear power.

5. Gallup obtained particularly low support for nuclear power by using a question with a clear antinuclear loading: "Do you feel that nuclear power plants operating today are safe enough with the present safety regulations, or do you feel that their operations should be cut back until more strict regulations can be put into practice?" In June, 1976, 34 percent responded "safe

enough," 40 percent said "cut back," and the remainder were undecided. In early April, 1979, shortly after Three Mile Island, those who thought the plants safe enough dropped to 24 percent, while those saying "cut back" rose to 66 percent; however by January, 1980, the "safe enoughs'" had risen back to 30 percent while the "cut backs'" dropped to 55 percent.

6. Q is a particularly convenient statistic for comparing crosstabular relationships from disparate sources because it is easily calculated from the crosstab, even in the absence of raw frequency counts. It is insensitive to the variations in the marginal distributions of variables which often occur across disparate polls. The Qs reported here are based on opinionated respondents only. The variables "age" and "education" were dichotomized near their medians.

7. Elsewhere I have reviewed five studies which test this explanation among the general public by measuring knowledge about the technology (using test questions) as well as attitude for or against it. The results are mixed, with two studies giving evidence that ignorance weakly predicts unfavorable attitudes, and three studies showing no relationship (Mazur, 1975). We may conclude that if ignorance explains any opposition among the general public, it is not much. These studies offer no support for the occasional opponent's claim that *support* of the technology is due to ignorance.

# 5
# Beliefs, Ideology, and Rhetoric

I want to make a distinct separation between our emotional *alignment* toward an object—our feelings for it of like or dislike, attraction, or repulsion—and our verbalized *rationale,* our set of articulated reasons, for that alignment.

Normally our alignments and rationales fit together. If we dislike fluoridation then we can usually state good reasons *why* we dislike it; if we dislike the people who oppose fluoridation, we can supply good reasons for that feeling as well. Occasionally alignment and rationale do not fit together, as when an environmentalist accepts the antifluoridationist warning against placing toxic material into the drinking water, and yet cannot feel much sympathy for the radical right-wingers whom he imagines to be at the center of the fluoridation protest. Such exceptions notwithstanding, it is usually the case that our alignment in a controversy has a good rationale which we can articulate when we are asked about it.

A core problem in the sociological study of controversy is to explain why people choose one side or another. If you *ask* people why they favor or oppose some technology, they will tell you, which means that they will state their rationale for the alignment. Thus, they favor nuclear power *because* it is the only new energy source that is realistically available in the next decades to keep our economy running; or they oppose nuclear power *because* there is no safe method for permanent disposal of radioactive waste products.

From the viewpoint of the partisan, his rationale may be a perfectly good explanation for his alignment. However, from the viewpoint of the sociologist, the rationale may be a satisfactory explanation or not, depending on whether it makes clear to us why that particular partisan took that particular position rather than another one. If a person said that he opposed the construction of an airport be-

55

cause his home lies in the proposed flight path and he objected to the noise that would result, then we would find this a satisfactory explanation because we all agree that life in a flight path is noxious, and anyone living there would likely become an opponent. However, if a biologist said that he opposed research using recombinant DNA techniques because of the great danger of creating virulent new organisms, then we might find this explanation incomplete because we know that many other biologists, equally concerned and informed, regard the danger of creating new organisms to be minor and easily guarded against (Grobstein, 1977). The rationale here seems as subjective as the position it is meant to explain, so we do not find it compelling. We will not be satisfied that we have a complete explanation of this partisan's position until we understand why he believes that the recombinant DNA hazard is great while others do not. His explanation only shifts our inquiry from the cause for his alignment to the cause for his rationale. In contrast, the rationale for airport opposition puts our inquiry to rest because it allows us to trace the route from the partisan's initial situation (living in the proposed flight path) to his eventual alignment, in easily understood steps.

If a partisan's rationale is the cause of his alignment, then we would expect that he must have learned of the rationale, and become concerned about it, before he became aligned. (If A causes B, then A must precede B.) However, there are numerous instances when rationale *follows* alignment. The most frequently heard reason today for favoring nuclear power is that it would reduce American dependence on Arab oil, a rationale that was virtually unknown before the Arab oil embargo of 1973, yet many of today's active proponents of nuclear power, mainly those in the nuclear industry, were proponents well before that date. When I interviewed leaders of the antinuclear movement in 1973, the most often stated reason for opposing nuclear power was that the emergency core cooling system for nuclear reactors had not been demonstrated to be effective on the scale of commercial power reactors. (This critical safety system would flood a reactor core with water in case the reactor lost its primary coolant, thus preventing a meltdown.) This problem had become public only within the previous two years (Gillette, 1971), at a time when most of these leaders were already active in the movement.

Another expectation we might hold, if rationale and alignment worked as simple cause and effect, is that once the rationale became known, then alignment would follow from it in a reasonably short time. The major problems of nuclear power have been known and publicized for years, and an active antinuclear movement has served since the late 1960's as a ready channel to facilitate the expression of opposition sentiments, yet large numbers of activists, particularly those closely identified with the anti-Vietnam movement, only entered the nuclear controversy after 1976 when the Clamshell Alliance focused national attention on the Seabrook power plant in New Hampshire. The Clam was soon joined by a new Crabshell Alliance in Washington, D.C., an Abalone Alliance in California, and others. Presumably, these people, at least those over 25 years old, could have entered the movement earlier if their opposition was a simple response to well known problems such as accidental radiation release, or long term storage of nuclear waste. However, it was only after 1976, when there was a dearth of liberal movements, that many perceived in the antinuclear movement a suitable outlet for protest.

An alignment, once formed, usually remains stable, though the supporting rationale may change over time. The reasons given to support a position for or against nuclear power have been in continual flux over the years of the controversy, but individuals rarely switch from a position in favor of the atom to one opposing it, or vice versa. Even an event as jarring to the rationale for nuclear power as the accident at Three Mile Island, produced little desertion of nuclear proponents to the opposition (Mazur, 1981a). In many cases one's alignment serves as an anchor point around which to interpret the various issues in the controversy, so that they may be incorporated into one's viewpoint in a consistent manner. Thus, to opponents of nuclear power, Three Mile Island demonstrated what they had argued all along: that serious accidents *will* happen, and that in this case we came very near to disaster. To proponents of nuclear power, the essential feature of Three Mile Island is that no one was killed, and that in spite of all sorts of errors, the safety system worked.

## Social Influence

When a partisan's stated rationale is not the cause of his alignment, what is? The best available answer is the one that permeates social psychology: We form our attitudes toward technology in the same way we form attitudes toward politics, religion, or any other topic—via social influence from friends, family, fellow workers, and public figures whom we respect (Gamson, 1966; Crain, 1966; Crain, et al., 1969; Mazur, 1975; Duncan, 1978). We tend to agree with those we like and associate with, and we disagree with those whom we dis-

whom we dislike or disapprove of. There is more to attitude formation than that, of course, but the main determinant is social influence.

When I graduated from college in 1961 with a bachelor's degree in engineering, I moved to California to take a job with a large aerospace company which manufactured intercontinental ballistic missiles. I found my friends and associates and my livelihood at that company, and lived and worked in a social milieu where nobody objected to building ICBMs, and neither did I. By 1969 I had left engineering and become a professor of sociology. Now I drew my friends, associates, and livelihood from the university of the 1960's, which had no sympathy for ICBMs, and neither did I by then. My attitudes were fairly typical of those around me because most of us reflect the influences of our social settings and of those people whom we consider to be like ourselves. Engineers I knew who were still in aerospace still had no objection to building ICBMs.

Several studies of recent social-protest movements indicate that recruitment often occurs along preexisting social links, and that people frequently join in an organizational bloc rather than as isolated individuals (Marx and Wood, 1975). The 30 leaders of the antinuclear movement whom I interviewed in 1973 showed this tendency. Two-thirds of these respondents had been active in the Environmental Movement prior to their opposition to nuclear power plants, their concerns about nuclear power developing largely in the context of local environmental groups which eventually affiliated with the antinuclear coalition. Half of these environmentalists explicitly reported that their concerns had been influenced by other antinuclear people with whom they had come into personal contact. A few nonenvironmentalists were introduced into antinuclear groups through friends, either as a social activity or in search of employment in the organization; their opposition developed only after becoming members of the group. Only a few respondents formed their antinuclear alignment completely independently of important social influences, three becoming involved because their homes were in immediate proximity to a proposed power plant or its transmission lines, and a few others saying that their antinuclear posture developed from literature which was critical of nuclear power.

The effect of social influence on alignment is apparent in the post-1976 phase of the nuclear power controversy. Prior to 1976, antinuclear activists were political liberals, but usually not those associated with the radical causes of the 1960's. Since the formation of the Clamshell Alliance in 1976, many people closely identified with these causes have joined the nuclear opposition, for example Jane Fonda and Tom Hayden. Thus, in the last few years, the antinuclear movement has become associated in people's minds with the student movement of the 1960's. Many people who were not particularly interested in nuclear power identified themselves with the student movement of the sixties. They now see the antinuclear movement populated by their kind of person, and they naturally feel some sympathy for it. If they see an antinuclear demonstration on television, reminiscent of the civil rights and anti-Vietnam demonstrations of the previous decade, they sympathize with the demonstrators. If the demonstration is nearby, they attend, particularly if some of their friends are going too (Van Liere, et al., 1979). Their enthusiasm and involvement grow, and they join the movement. Along the way they become knowledgeable about the problems of nuclear power, and they develop rationales to explain their opposition to it.

A similar, though less intense, effect is occurring on the other side of the controversy, where many who disapproved of the student radicals during the sixties now make facetious remarks about Jane Fonda, hope that antinuclear demonstrators will be arrested, defend nuclear power, and, of course, form rationales to support these positions.

A word needs to be said here about the respectability of social influence. Activists who have read the above discussion may feel that it ignores the individual's own inherent moral sense as a determinant of his behavior, treating him like a sheep whose actions are predetermined by the movement of the flock. It is not my intention to demean activists or their actions, and I apply the effects of social influence as readily to explain my own behavior as to explain the work of an activist. Everyone is conditioned and affected by the social influences which currently surround him, and by those which have surrounded him in the past—that is the nature of a social being. Some people will conform more quickly than others to the expectations of the immediate social setting, for example, the young person whose antinuclear protests begin and end within a few weeks of a large rock-music demonstration which he attended, but I have explicitly excluded such ephemeral behavior from activism, focusing instead on those who are persistently concerned.

However the fact that a concern is persistent does not mean that it was formed and sustained independently of social influences. Many who were on the university campuses of the 1960's were deeply concerned about black civil rights and acted on those concerns, much more so than those on the same campuses in the decades before and after. We have no reason to believe that the students and faculty of the 1960's were better people, more sincere and humane than those of the 1950's or 1970's, so why were their actions and concerns so different? It cannot be that the problem was more pressing then because the needs of the nation's poor blacks were about as great in the 1950's and 1970's as in the 1960's. What was different was the social milieu of the campuses of the 1960's, which had begun with the idealism of the Kennedy presidency. If one were in an academic sociology department at that time, the civil libertarian influence was strongly felt, as was the anti-Vietnam influence a few years later. Nearly everyone was involved, including people who had never before expressed much interest in racial equality, and many stayed involved throughout the decade, their sincerity unquestioned. Yet there is no doubt, in retrospect, that their actions, and often concerns, were produced by the prevailing social influences of that setting at that time, which were not there previously or afterward.

## The Content of Rhetoric

As the lines are drawn in a technical controversy and the positions become polarized, each side develops a rhetoric, an articulated ideology that is more or less shared by partisans. This rhetoric includes the individual rationales which are most popular as well as other beliefs. New recruits to each side of a controversy are socialized into those parts of its ideology which they do not hold already (Gerlach and Hine, 1970).

The degree of attitude consensus on either side of a controversy is often overstated, particularly by detractors who stereotype their opponents. Usually each side has topics of internal dissension which are as marked as their points of agreement, though the dissent may not be apparent to an outsider because of the partisan tendency to present a solid front to anyone not in the group, particularly to an adversary. Perhaps the most common area of disagreement is tactical decision making about protest or promotional activities. Tactical disputes have been particularly devisive among antinuclear protesters, for example, whether or not to use violent or illegal means to occupy nuclear facilities and to disrupt their operations (List and Katz, 1979; Barkan, 1979). Such disputes may encourage the formation of factions within a movement, or they may reflect an already existing factionalism. The protests against high-voltage electric transmission lines in upstate New York and Minnesota were carried out by coalitions of conservative farmers and liberal young people, often students. Not surprisingly, the groups differed on social issues such as abortion and women's rights; since these issues were not relevant to the transmission line dispute, however, the farmers and liberals managed an accommodation, primarily by excluding such items from conversation, focusing instead on topics of closer agreement.

Each side of each controversy has its own unique ideology; the rhetoric in one controversy, however, is often strikingly similar to rhetoric from another controversy, as illustrated in Chapter 2 where I gave several pairs of nearly interchangeable statements from the nuclear power and fluoridation controversies. Sometimes the similarities come from a direct borrowing, particularly when participants from one controversy enter another, bringing along their established beliefs and rhetoric. Thus, the "Hell no, we won't go," of the Vietnam era becomes "Hell no, we won't glow," when adapted to nuclear power. However, this sort of carryover is not an appealing explanation of similarities between fluoridation and nuclear power because these two movements had different personnel, and there was little sympathy or communication between them. What they had in common was a particular "structure."

Social structural features of a controversy can shape the content of rhetoric in explicit ways, so if two controversies have similar structures, they may show rhetorical similarities as well. When one side promotes a technology which the other side opposes, the two sides form differing beliefs about the costs and benefits of the technology, which become part of their respective ideologies. The side that represents corporate interests will develop an establishmentarian rhetoric which will be countered in predictable ways by the challengers. These account for some of the similarities in the rhetorics of fluoridation and nuclear power. Furthermore, there are similarities in people's perceptions of radiation and fluoridation, each being an invisible hazard introduced into the environment. Thus, there are ample reasons to expect rhetorical similarities across controversies even in the absence of any diffusion from one to the other.

Three principles explain some of the rhetorical content of technical controversies. The first two principles relate social structure asymmetries to content, the two asymmetries being that one side favors the technology while the other side doesn't, and that one side represents the establishment while the other side doesn't. The third principle relates rhetorical content to the nature of the technology.

**Proponents versus Opponents**

Each controversy has one side favoring a technology and another side opposing it, a situation which carries certain imperatives for the formation of rhetoric. Obviously, the proponents will require an ideology which praises and promotes the technology, while opponents need an ideology of criticism and condemnation.

The proponents of radiotherapy for "enlarged" thymus were devoted to three central beliefs: that there was an important *need* for the therapy, that the therapy was *effective,* and the therapy was *safe* (Chapter 1). The same three beliefs—need, effectiveness, safety—are central to the proponent ideology behind every technology that has been a focus of controversy. (Example: Tooth decay is rampant in American children; fluoridation prevents cavities without ill effects.) These beliefs are the core of the promotion: "You have a problem; I can solve it; the solution won't hurt."

Opponents necessarily disagree, usually on all three counts, by minimizing the problem, by questioning the efficacy of the proposed technology (compared to alternate solutions), and by showing that the proposed technology is hazardous. (Example: Cavities are a minor affliction; fluoridation simply delays cavities, it doesn't prevent them, and we would do better by removing sugar from children's diets; fluoride is poison.) Studies of beliefs about nuclear power consistently show that proponents, compared to opponents, see a greater need for nuclear power, greater benefits flowing from it, and smaller risks (Harris and Associates, 1976; Stallen and Meertens, undated; Maynard, et al., 1976).

Since so many controversies have similar rhetoric for or against the technology, some observers have suggested that these many disputes are all facets of a single megacontroversy pitting those who believe in progress through science and technology against those with a broad antiscience-and-technology ideology. The inadequacy of this view is demonstrated by noting, first, that many partisans and even leaders of particular antitechnology movements are themselves eminent and productive scientists. Furthermore, we often find specific individuals among the opposition to a technology who are strong proponents of some other technology. Nobel physicist Hans Bethe was a major opponent of the ABM and is a major promoter of nuclear power. Barry Commoner is a strong opponent of nuclear power but a major proponent of photovoltaic solar technology (Commoner, 1979). Nobel biologist James Watson, another opponent of nuclear power, favors recombinant-DNA research. Ralph Nader opposes nuclear power and promotes state-of-the-art technology for auto safety. These men object to particular aspects of particular technologies; they do not oppose science and technology as a general principle. One need not assume that all controversies with similar ideology are linked together. The structural similarity among them, one side always favoring and the other side opposing a technology, can account for rhetorical similarities.

## Challengers versus the Establishment

Another important structural feature of most controversies is that one side is associated with an establishment position representing corporate business, the professions, or the federal government. The other side sees itself as challenging this mighty establishment and naturally develops a rhetoric which emphasizes the virtue of the common man standing up for his rights against big business, big government, or the military-industrial complex. Challenge groups often refer to themselves as "citizen," "consumer," or "public interest" associations, suggesting that they speak for the average person. They see the establishment side as very powerful, capable of suppressing information or manipulating the mass media through bribery or coercion. Almost invariably the establishment is seen as forcibly imposing its will on the populace, violating their rights as citizens, and perhaps endangering their lives in the process. If the challengers are political leftists, as in the ABM or nuclear power controversies, then the establishment may represent corporate capitalism, grasping for profits. If the challengers are political rightists, as in the fluoridation and Laetrile controversies, then the establishment may represent creeping socialism, smothering the rights of the individual. In either case, some ideologists see a conspiracy at the center of the establishment position, the plotting of a few ambitious and powerful men who manipulate the society for their own advantage.

The establishment's rhetoric emphasizes the nonorthodox position of the challengers, portraying them as deviant (hence disreputable), uninformed and perhaps misguided (if well meaning). When challengers lack proper academic or professional certification as experts, this is used to undermine their credibility; when they are properly certified, they are attacked for bringing their case to the mass media rather than publishing in peer-reviewed professional journals; when they do present their cases in professional journals and meetings, establishment experts prepare refutations showing bias and error in the analysis. The challengers are seen as naive, irrational, disruptive violaters of proper procedure, extremist, alarmist, either radical rightists or radical leftists (depending on the right or left orientation of the challengers), fanatic, myopic, and stupid but crafty. Some establishmentarians believe that a single conspiratorial network unites all challenge movements, and that it is financed from sources outside of the country.

In most controversies the establishment side is the promoter of the technology, and it therefore appraises need and benefit from the perspective of corporate, federal America. Thus, it is believed that the technology will spur the economy, or it will improve national defense. On the other side, the challenger-opponents of the technology believe that it is being forced on the people, to the detriment of both the populace and the environment, in order to enhance the power of the elite.

When the establishment *opposes* the technology, as in the controversy over the reputed cancer cure, Laetrile, then it views the risks of adoption to be grave for the nation, and the benefits to be dubious, largely because of the unorthodox credentials (hence questionable skill and motives) of practitioners. The unorthodox challenger-promoters believe that they are being persecuted and that their individual rights are being violated because they threaten the prerogatives of orthodox practitioners.

Similarities in antiestablishment rhetoric across controversies, like similarities in antitechnology rhetoric, have given rise to a theory of megacontroversy which links all of the challengers in these diverse disputes into a single coalition (The Movement) which is fighting an opposing coalition of all the diverse establishment partisans (The Establishment). The fact that this theory of massive confrontation exists in the minds of some partisans gives it some reality—at least symbolic reality—but I believe that its importance has been exaggerated. There are many instances when partisans in a controversy become antiestablishment only after the fact. A protest over a particular power line, or an airport, or a nuclear power plant may begin when local residents perceive that the new project will intrude on their property or present a hazard. In this sort of situation, when common citizens must take their complaints to electric utility corporations and government agencies, and when they do not receive satisfaction, it is natural to develop a viewpoint which emphasizes the problems of the little man in his battle with the larger powers.

Undoubtedly there are people who sometimes join protests because of their ideological objection to establishment control. If this were a deeply felt motive, then any movement which protests against establishment control should be attractive to such people. However most challenge partisans are quite selective in supporting protests. While antifluoridationists receive support from pro-Laetrilists (Markle, et al. 1978), neither of these conservative antiestablishment groups gives or receives support from the protest groups of the left. I have often raised the subject of fluoridation in discussions with liberal environmentalists and antinuclear activists, and while they might agree that the addition of fluorides to a community's water supply is an environmental insult, only a few support the protest against fluoridation. In many cases they were clearly reluctant to associate themselves with "right-wingers." Perhaps if fluoridation had a protest from the left rather than the right, then many of these people would be active supporters of that movement. One's left-right political alignment seems at least as important a motive for joining a protest as one's particular ideology regarding the establishment. In the following chapter, when I describe the movement of elite scientists from one controversial issue to another, we will see additional deviations from pure pro or antiestablishment positions, as when scientists in John Kennedy's circle joined the "antiestablishment" protests against the SST and the ABM, and when opponents of the ABM joined the "proestablishment" supporters of nuclear power plants. These alignments are not consistent with the notion of a solid front dividing the establishment from its challengers.

## Classes of Technology

So far I have pointed to ideological content which arises as a natural consequence of asymmetries between the two sides, one side being the establishment while the other challenges the establishment, and one side promoting the technology while the other side opposes it. Thus much of the content of the controversy comes from these social structural features and is independent of the particular technology that is in dispute.

Obviously the rhetorical content will also be shaped to fit the specifics of the technology, for example, beliefs about whether or not supersonic transports will deplete the earth's protective ozone layer. However, each technology need not be treated purely on its own terms but instead may be regarded as one example of a particular class of technologies. There is, for example, the important class of technologies which give people low doses of something which, in higher doses, is very toxic. These substances include fluoride, radiation, saccharin, some pesticides, and certain food additives. Controversies over technologies in this class almost always require ideological positions on, say, the proper form of dose-response curves at very low dosages. Typically such curves are fairly well specified at higher doses but not at the lower levels typical of population exposure. Should one assume a linear dose-response relationship, or is a threshhold effect more appropriate (Chapter 2)? If cancer is found in rats given large doses of the substance in a laboratory experiment, what inference should be drawn about cancer induction in human populations receiving chronic low doses? How firm must such evidence be to justify regulatory action by government agencies? How many cancers should we tolerate in exchange for the benefits of the technology? These issues appear repeatedly, and the positions of opponents and proponents are predictable.

Another important class consists of those technologies which have the potential for very low-probability but very high-consequence accidents. Examples are nuclear power plants, the ABM, and recombinant DNA. Proponents invariably emphasize the very low probability of accidents, while opponents focus on the very high consequences.

Any technology that is very costly will produce ideological positions on corporate profit and responsibility. Any military technology will require positions on foreign policy. A technology that is deployed at specific sites will produce positions on local (community or state) versus distant (state or federal) control. Any technology which cannot be adopted or rejected voluntarily will produce positions on individual rights and freedom of choice. Thus, it is possible to predict several features of ideology from the general characteristics of the technology of concern, without regard for its specifics.

We have seen that many of the ideological disputes in a controversy arise as natural consequences of structural asymmetries between the two sides; such disputes occur independently of the particular technology that is in question. Also, many partisans articulate the reasons for their alignments only after they have chosen a side, often on the basis of social influence. Thus, from the sociologist's perspective, rhetoric and ideology cannot be accepted at face value as the reason for a controversy or as the motive for a partisan's involvement. There are surely instances when such statements of belief reveal the core of a dispute, but there are other instances when they are merely epiphenomena.

I suggest that the most important links between controversies are not their ideological similarities, which may only be surface features, but rather the social networks which tie some activists together. The formation of coalitions and the pooling of resources are based more on these networks than on shared ideologies. These social links, particularly the sharing of personnel between one controversy and another, are explored further in the next chapter.

# 6
# Social Links Among Controversies

*(with George Misner)*

No single group of activists spans the whole set of contemporary technical protests. The antinuclear critics on the political left have little to do with the corporate opponents of the airbag, and neither of these is linked to the grassroots conservative antifluoridationists or to the anti-Laetrile leaders from the nation's biomedical establishment.

The most visible sharing of personnel among controversies occurs in the academic science community. University scientists have been prominent in many American political controversies and citizen movements of the past twenty years, some of these having important technical components (e.g., the dispute over the supersonic transport) and some being essentially nontechnical (e.g., the Peace Movement). These controversies are an integral part of the politicization of American science since World War II. In this chapter I will describe social links among them.

## Controversies in Academia

In most ways scientists are like other people. They tend to choose friends and associates who are like minded, and if not initially like minded, then friends' attitudes and beliefs become more similar with the passage of time (though everyone knows exceptions). Anyone who joins a controversy will usually have friends on the same side, either because he brought them in, or they brought him in, or both enlisted because of their similar circumstances. Someone moving to a new controversy is usually accompanied by friends from the last controversy, for the same reasons. Thus, a convenient way to examine links among controversies is to trace the people who move from one to another. Controversies are linked or not, in this sense, depending on whether or not they share a relatively large number of personnel.

In practice, it is not easy to trace out migration from one controversy to another because there are no complete censuses of the individuals on each side of the controversies of interest. The situation is not hopeless because adequate (if imperfect) lists do exist for a number of issues, and these are the basis of the following analysis.

I began by collecting 14 lists of names of scientists who had taken a particular position on specific issues in American politics between 1960 and 1976. Eight lists come from petitions published in *The Sunday New York Times,* which has been the most frequent vehicle for political petitions from the academic community (Ladd, 1969, 1970; McCaughey, 1976). Since my primary interest is in scientists, broadly defined to include technologists and academic physicians, I used only petitions where signers could be identified by discipline, and scientists' names were taken for the lists. The first *Times* petition supported the presidential candidacy of John Kennedy in 1960, and I obtained similar lists supporting the Johnson-Humphrey ticket of 1964 and the peace candidacy of Eugene McCarthy in 1968. President Kennedy's national fallout shelter program was attacked in a series of petitions in 1961 by scientists who opposed escalating the Cold War with Russia. The largest series of *Times* petitions attacked President Johnson's escalation of the Vietnam War, though my list is limited to the early petitions (1965) when opposition to the War was still a relatively radical position in the academic community. The following year, a small petition by prestigious scientists appealed to President Johnson to stop the use of chemical weapons in Vietnam. I also found a 1970 *Times* petition opposing construction of Egyptian missile sites in violation of the cease fire agreement with Israel. None of these were overtly technical issues, but I examined three other issues which had scientific disputes at their cores: the antiballistic missile, the supersonic transport (SST), and nuclear power plants.

A large *Times* petition opposed deployment of the "Safeguard" ABM system in 1969, claiming that it would provide little protection for the United States, would escalate the arms race, would hinder disarmament negotiations, and would consume funds that ought to meet urgent civilian needs. This petition formed one of three lists I obtained for the ABM controversy. Two others were provided by Cahn (1974), who had interviewed scientists active on both sides of the ABM dispute; one list contained her pro-ABM subjects and the other her anti-ABM subjects. Lists of scientists for and against the SST were obtained from the *Congressional Record* (1970, 116: 39777,

41594), supplemented by Primack and Von Hippel's (1974) account of that controversy. I obtained a list of antinuclear power names from a large petition circulated in 1975 by the Union of Concerned Scientists of Cambridge, Massachusetts, and I compiled a pronuclear power list from two petitions circulated in response to the opposition petition (Boffey, 1976). The 14 lists and short mnemonic titles (in capitals) appear in Exhibit 10.

My lists have a number of problems which restrict the conclusions which may be drawn from them: (1) I could not include all of the political issues which have involved the scientific community; (2) heavy dependence on newspaper advertisements overemphasizes one type of political activity—signing ads,—and it is not clear what sort of commitment is implied by that gesture; and (3) *The New York Times,* with its focus on the East Coast, gives inordinate emphasis to that area's political activity. The lists do not randomly sample American scientists; the urban Northeast is surely overrepresented.

The ideal data set for this study would be a survey of various political activities taken from a properly drawn sample of American scientists who were interviewed several times over the past two decades. Unfortunately, I have no way to reach that ideal. The lists used here are nearly all of such lists which are in existence, so far as I know. A few additional *Times* lists are either redundant, or very small in size, or do not permit specification of signers by scientific discipline. The availability of Cahn's ABM lists, and of the petitions from the nuclear power and SST controversies are fortuitous. While these lists do not cover all of the political issues of the scientific community, they do illustrate the range of these issues and include those which have been most salient. The appearance or not of a given name on a given list probably depends as much on the personal acquaintances of the list organizers as any other fact, but of course, such social networks are precisely the subject of this study. Furthermore, I have checked each list against my broader knowledge of each controversy and am satisfied that in all cases, nearly all of the principal scientists who were involved on a side appear on an appropriate list representing that side. These lists are the best series data on political activity of scientists which are available, and are adequate for the present purposes, as long as reasonable caution is used in drawing inferences.

Of about 7,500 scientists appearing on at least 1 list, I located 367 (5 percent) who were on 2 or more lists.[1] Most of the analysis here is based on these multiply listed names, so I will begin by examining how they differ from the bulk of names which appear on only one list.

**EXHIBIT 10**

Lists of Scientists Associated with Specific Political Positions.

| Mnemonic Label | Political Position | Size of List | Source of List | Date(s) |
|---|---|---|---|---|
| KENNEDY | Support John Kennedy for President | 151 | *New York Times* | 10/30/60 |
| SHELTER | Oppose President Kennedy's fallout shelter program; de-escalate arms race | 765 | *New York Times* | 11/10/61<br>12/19/61<br>2/18/62 |
| JOHNSON | Support Johnson-Humphrey presidential ticket | 215 | *New York Times* | 10/11/64 |
| VIETNAM | End the war in Vietnam | 944 | *New York Times* | 2/28/65<br>3/28/65<br>4/11/65<br>5/9/65 |
| CHEMWAR | Halt American use of chemical weapons in Vietnam | 22 | *New York Times* | 9/20/66 |
| MCCARTHY | Support Eugene McCarthy for President | 218 | *New York Times* | 3/17/68<br>3/21/68<br>6/23/68 |
| ABM-NYT | Oppose deployment of the "Safeguard" ABM | 1800 | *New York Times* | 3/23/69 |
| ANTIABM | Oppose deployment of an ABM | 106 | Cahn (1974) | 1969 |
| PROABM | Support deployment of an ABM | 40 | Cahn (1974) | 1969 |
| ISRAEL | Oppose Egypt's violation of cease fire agreement with Israel | 26 | *New York Times* | 9/20/70 |
| ANTISST | Oppose the supersonic transport | 30 | *Congressional Record* | 1970 |
| PROSST | Support the supersonic transport | 34 | *Congressional Record* and Primack & Von Hippel (1974) | 12/15/70 |
| ANTINUCLEAR | Oppose nuclear power plants | 2000 | Petition by Union of Concerned Scientists | 8/6/75 |
| PRONUCLEAR | Support nuclear power plants | 155 | Atomic Industrial Forum's *INFO*, and "Scientists' Statement on Safety of Nuclear Energy" (mimeo) | 2/76<br>5/1/76 |

## Multiply Listed Scientists

For most multiply listed names, and for a random sample of 100 singly listed names, I was able to determine scientific discipline, geographic location, sex, membership in the National Academies of Sciences and Engineering or the Institute of Medicine (hereafter referred to as NAS), receipt of a Nobel Prize, and membership in the Federation of American Scientists(FAS), an organization of politically active scientists (Nichols, 1974).[2] Exhibit 11 shows these characteristics as a function of the number of lists upon which a person's name appears.

As the number of listings increases, we find higher proportions of prestigious scientists (as measured by NAS membership and receipt of a Nobel Prize), higher proportions of FAS membership (which, like number of listings, is an indicator of political activism), higher proportions from the greater Boston area universities along with decreasing proportions from outside the Northeast, and higher proportions of physicists and slightly more biomedical scientists along with decreasing proportions of social scientists.

Activist scientists are often pictured as political leftists, with notable exceptions. The data support this tendency, but one should be wary of oversimplification here. The American academic community is predominately liberal/left/Democrat compared to the rest of the nation, particularly at the most prestigious universities (Ladd and Lipset, 1972, 1976). In this context the virtual absence of conservative movements, say in support of the Goldwater presidential campaign, is to be expected.

Issues such as support for Israel or the Johnson presidential campaign did not have an explicit left-right polarization, so activism in these cases does not indicate radical politics. Certain lists do indicate a clear left orientation relative to the academic community, especially SHELTER, VIETNAM, and MCCARTHY. On the other hand, three lists seem relatively conservative: PROABM, PROSST, and PRONUCLEAR (Cahn, 1974; Gray, 1975; Mazur, 1975). The remaining lists fit between these extremes. Exhibit 11 shows the percentage of scientists appearing on each list as a function of number of listings. (For example, of those who appear on three lists, 66 percent appear on SHELTER.) Those most often listed were much more likely than singly listed people to appear on the "left" lists, but they showed only a slight rise in representation in the conservative lists.

In sum, the multiply listed, compared to the singly listed, are more likely to be prestigious, to be physicists or biomedical scientists,

EXHIBIT 11

Characteristics of Scientists by Number of Listings

| | | Number of Listings | | |
|---|---|---|---|---|
| | 1*<br>(n = 6100) | 2<br>(n = 294) | 3<br>(n = 55) | 4-6<br>(n = 18) |
| Nobel Laureates | 0 | 5 | 16 | 17 |
| In NAS | 2 | 23 | 51 | 56 |
| In FAS | — | 7 | 24 | 33 |
| *Academic Field* | | | | |
| Physics | 13 | 21 | 22 | 33 |
| Biomedical | 28 | 36 | 33 | 33 |
| Social science | 28 | 18 | 22 | 0 |
| Other | 30 | 24 | 24 | 33 |
| | 99 | 99 | 101 | 99 |
| *Geographic Location* | | | | |
| Boston area schools | 10 | 28 | 35 | 61 |
| New York City area schools | 12 | 45 | 42 | 28 |
| California schools | 7 | 4 | 6 | 6 |
| Other | 70 | 22 | 18 | 6 |
| | 99 | 99 | 101 | 101 |
| Female | 9 | 10 | 9 | 0 |
| *On Each List* | | | | |
| SHELTER | 10 | 44 | 66 | 72 |
| VIETNAM  Left issues | 13 | 40 | 53 | 50 |
| MCCARTHY | 3 | 10 | 27 | 44 |
| KENNEDY | 2 | 7 | 16 | 33 |
| CHEMWAR | 0 | 1 | 15 | 17 |
| ANTIABM | 1 | 8 | 9 | 22 |
| ABM-NYT | 29 | 40 | 40 | 39 |
| ANTISST | 0 | 2 | 4 | 17 |
| ANTINUCLEAR | 31 | 24 | 27 | 61 |
| JOHNSON | 3 | 16 | 24 | 28 |
| ISRAEL | 0 | 2 | 4 | 17 |
| PROABM  Conserva- | 1 | 1 | 2 | 6 |
| PRONUCLEAR  tive | 2 | 5 | 11 | 11 |
| PROSST  issues | 1 | 0 | 4 | 6 |

Percentage

* Based on a random sample of 100.

to be active on the political left, and to come from the greater Boston academic community. The two most frequently listed scientists, Harvard biologists George Wald and John Edsall, fit this pattern.

## List Categories

One can trace the links between lists, looking in particular at whether or not the various protechnology lists (PROABM, PROSST, and PRONUCLEAR) form one cluster while the antitechnology lists (ANTIABM, ABM-NYT, ANTISST, and ANTINUCLEAR) form a separate cluster. However, before moving to that examination, it is convenient to sort the 14 diverse lists into 3 categories which differ on some readily observable characteristics.

As the lists decrease in size, they tend to increase in prestige (measured by proportion of NAS members or of Nobel laureates) and in the proportion of political activists (measured by proportion of multiply listed names), as illustrated in the log-log plot of Exhibit 12. The covariation of size, prestige, and activism is easily explained by the tendency, shown above, for activists to be prestigious, plus the obvious futility of publishing a small petition in *The New York Times* unless the signers were of recognized importance. However, these three variables are not perfectly correlated, and I have marked three regions on the log-log plot which are conveniently regarded separately. The lower region contains the four largest lists which also have the lowest proportions of NAS members and are low to moderate in proportion of multiply listed activists. I regard these as *popular* lists while the other ten, in the upper half of the plot, are small *elite* lists with high proportions of prestigious scientists. The ten elite lists are further subdivided into a *protechnology* category (PRONUCLEAR, PROABM, and PROSST) with relatively few persistent activists, and the remaining seven lists (to the upper right on the plot) with high proportions of activists.

These three categories of lists may be characterized further using their geographic and discipline distributions of *multiply listed* scientists, shown in Exhibit 13. (Multiply listed scientists constitute biased samples of each list but are appropriate given my concern with persistent activists.)

The four popular lists in the bottom region of the log-log plot are drawn primarily from the urban Northeast, particularly the greater Boston and New York City areas. This geographic focus no doubt reflects, to some degree, the fact that three petitions come from *The New York Times* (SHELTER, VIETNAM, and ABM-NYT) while the fourth (ANTINUCLEAR) was organized by the Union of Concerned Scientists, based in Cambridge. A wide range of academic disciplines is covered in each of these lists. In sum, these four lists are drawn from the undifferentiated academic mass of the urban Northeast, and I will refer to them as "northeast popular" lists. Perhaps this large mass of signers moves from one petition to another.

In contrast, the three elite protechnology lists (PRONUCLEAR, PROABM, and PROSST) have minimal representation from Boston

EXHIBIT 12

Size, Percentage NAS Membership and Percentage Multiply Listed Scientists for all
Lists (Size of list appears in parentheses)

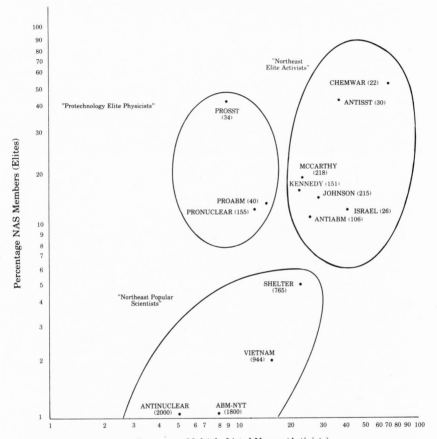

and New York, drawing largely from California, and their multiply
listed scientists are predominantly physicists. These people may form
a protechnology core, moving from one controversy to another.

The seven other elite lists cover a range of disciplines and, like
the popular lists, draw mainly from the Northeast, but more from
Harvard (JOHNSON is an exception). Perhaps these "northeast elite
activists" form another core, moving not only among antitechnology
lists but other political issues as well.

In order to check these conjectures, one may look directly at the
degree to which the same names appear on the various lists.

EXHIBIT 13

Geographic and Discipline Distributions of Multiply Listed Names on Each List

| | "Northeast Popular Scientists" | | | | "Protechnology Elite Physicists" | | | "Northeast Elite Activists" | | | | | | |
| --- | --- | --- | --- | --- | --- | --- | --- | --- | --- | --- | --- | --- | --- | --- |
| | SHELTER | VIETNAM | ABM-NYT | ANTINUCLEAR | PRONUCLEAR | PROABM | PROSST | KENNEDY | CHEMWAR | MCCARTHY | JOHNSON | ANTIABM | ANTISST | ISRAEL |
| n | (179) | (155) | (144) | (97) | (22) | (6) | (3) | (36) | (15) | (53) | (64) | (31) | (11) | (12) |
| *Geographic Location* | | | | | | | | | | | | | | |
| Harvard | 9 }29 | 7 }43 | 8 }38 | 12 }29 | 5 }10 | 0 }0 | 0 }0 | 30 }36 | 27 }40 | 29 }41 | 2 }7 | 13 }32 | 46 }64 | 42 }67 |
| Other Boston area schools | 20 | 36 | 30 | 17 | 5 | 0 | 0 | 6 | 13 | 12 | 5 | 19 | 18 | 25 |
| New York City schools | 66 | 47 | 43 | 34 | 18 | 17 | 0 | 15 | 20 | 22 | 63 | 3 | 0 | 8 |
| California schools | 0 | 0 | | 1 | 32 | 33 | 67 | 18 | 20 | 14 | 9 | 7 | 0 | 8 |
| Other | 5 | 10 | 19 | 37 | 41 | 50 | 33 | 30 | 20 | 22 | 22 | 58 | 36 | 17 |
| | 100 | 100 | 100 | 101 | 101 | 100 | 100 | 99 | 100 | 99 | 101 | 100 | 100 | 100 |
| *Discipline* | | | | | | | | | | | | | | |
| Physics | 12 | 21 | 18 | 25 | 55 | 67 | 67 | 33 | 27 | 28 | 19 | 55 | 18 | 8 |
| Biomedical | 50 | 23 | 29 | 53 | 5 | 0 | 0 | 22 | 47 | 43 | 58 | 7 | 9 | 0 |
| Social science | 19 | 28 | 23 | 5 | 0 | 0 | 0 | 22 | 0 | 2 | 3 | 10 | 36 | 67 |
| Other | 20 | 28 | 30 | 17 | 40 | 33 | 33 | 22 | 26 | 26 | 20 | 29 | 36 | 25 |
| | 101 | 100 | 100 | 100 | 100 | 100 | 100 | 99 | 100 | 99 | 100 | 101 | 99 | 100 |

Percentage

## Links Between Controversies

My pool of 367 multiply listed names will show the degree to which names on one list coincide with names on any other list. If two lists, A and B, are linked by a moderately large social network, then multiply listed names on List A should be more likely than those not on A to appear on List B. Put another way, if Lists A and B are linked, then the presence of a multiply listed name on A would increase the probability that it appears on List B. The commonality of names between two lists is conveniently measured by the correlation coefficient Q which takes values from $-1.0$ to $+1.0$, where positive values indicate commonality of names between lists, zero indicates that the lists are independent, and negative values indicate that names on List A tend to be excluded from List B. The value of Q relating any two lists has the desirable feature of being independent of the *number* of multiply listed names on either list; thus the size of lists does not affect Q.[3]

There are ambiguities in the interpretation of Q. A negative Q between Lists A and B might mean that signers of one list avoid signing the other list, or it might simply mean that signers of one list had no opportunity to sign the other, perhaps because the lists were circulated in different locales. Furthermore, a small network of activists might move from List A to List B and escape detection (Q being zero or even negative) if the bulk of "non-network" names changed from list to list. Another problem exists in the calculation of Q for small lists since a single common name might lead to a deceptively high value of Q. To avoid some of these problems, my interpretation is based primarily on positive Qs of .2 or greater, ignoring those due to a single common name. These provide a good indication of social links among lists.

The diagram in Exhibit 14 shows all lists in their proper time relationship. Pro and antitechnology lists are crosshatched to separate them visually from nontechnical lists. Elite lists are toward the top of the diagram and popular lists are shown at the bottom in larger boxes. Arrows (pointing in the direction of advancing time) link lists that are related by a positive Q (of .2 or greater); thicker arrows are associated with larger values of Q. In parenthesis, after the Q value, is the number of names which two lists have in common. (This number is necessarily small when the two lists are small.)

Three of the four popular lists at the bottom of the diagram are linked, substantiating the existence of a bloc of scientists moving

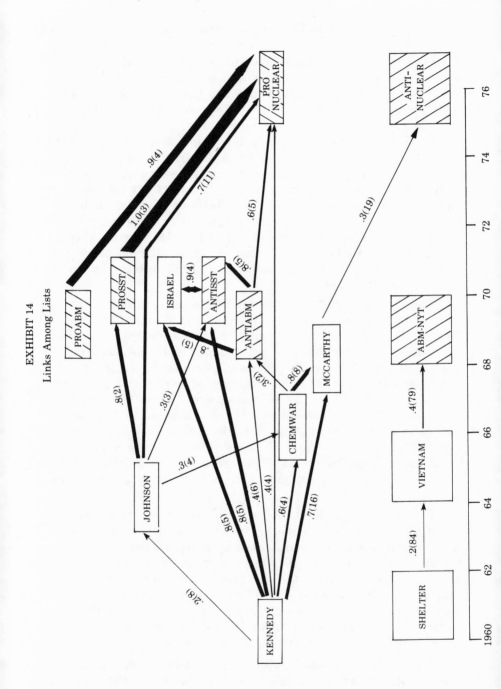

**EXHIBIT 14**
Links Among Lists

from one popular issue to another; however, the chain does not extend to ANTINUCLEAR, the most recent popular list.

Of three possible links among the three protechnology lists, two are strong, and there is further linkage to the JOHNSON list. However pro and antitechnology lists do not stand out as separate clusters. To the contrary, the overall impression is one of widespread interconnection among all the elite lists, even between ANTIABM and PRONUCLEAR.[4]

The striking bifurcation is between the elite lists, whether pro or antitechnology, and the popular lists. This is particularly dramatic in the ABM issue where there are two opposition lists, one elite (ANTI-ABM) and one popular (ABM-NYT). The Q between the lists is strongly negative ($-.9$, only one name appearing on both lists), though about one-third of each list's multiply listed names came from the greater Boston area (Exhibit 13). These two anti-ABM lists are nearly as isolated from one another as each is from the PROABM list.[5]

What differentiates scientists who remain exclusively in the elite lists from those who are confined to the popular lists, and how do these differ from scientists who cross the line, appearing in both popular and elite lists? One might expect that those who appear on the most lists would be most likely to move between the strata, but there is only a slight tendency in that direction. Scientific prestige is the major factor I could associate with strata behavior. Among multiply listed scientists, 60 percent of Nobelists appeared only in elite lists compared to 32 percent of non-Nobel NAS members compared to 6 percent of the rest; conversely only 12 percent of the Nobelists were exclusively on popular lists compared to 25 percent of non-Nobel NAS members compared to 75 percent of the rest. Thus, the more prestigious the scientist, the more he restricted himself to elite lists.

Why do multiply listed scientists tend to be bifurcated into elite and popular networks? It seems likely that this division reflects a real separation of political activity between a relatively small group of prestigious professors and the much larger body of middle and lower status scientists. That is not to say that these strata never work together or share common causes, because they obviously do. Nonetheless, it does appear that the political activity of the scientific elite is carried out in relative isolation from popular activity.

The maintenance of a stratified political system in science, and perhaps its origin, likely follows from the relatively closed social world of prestigious scientists whose careers, from student days onward, are bounded by the campuses of the nation's most prestigious universities, by their exclusive participation in councils of scientific advice to government, by memberships in the National Academy of Sciences, and particularly by the informal ties of respect and regard for mutual friends, as they interact at frequent personal meetings and over long distance telephone (Cole and Cole, 1973; Greenberg, 1967; Kash, White, Reuss, and Leo, 1972; Zuckerman, 1977).

This two-tier picture of scientific politics is consistent with historical events in the Boston academic community. John Kennedy's presidential campaign was heavily supported there, particularly at his alma mater, and many prestigious Harvard professors subsequently served the Kennedy administration. In 1961, a group of scientists from Harvard, MIT, Brandeis, and Boston University formed to protest Kennedy's Bay of Pigs invasion and, shortly thereafter, organized the *Times* petition against his fallout shelter program; some of this group were later active in petitioning against the Vietnam War (Ladd, 1969, 1970). Thus by 1961, if not earlier, Boston academicians were divided into a prestigious group centered at Harvard and committed to Kennedy, and a more broadly based protest group opposed to some of Kennedy's major policies. Exhibit 14 suggests movement of the elite Kennedy people to a series of issues compatible with the liberal Democratic politics of the time.

## The Nuclear Power Controversy

Abundant sociological research has shown participation in one or another social network to affect political (and other) attitudes. Social influences are particularly interesting in cases of technical policy where, it is often assumed, objective decisions are based on facts, though this assumption has been criticized for its naivete (Cahn, 1974; Gray, 1975; Mazur, 1973).

The nuclear power controversy, with its petition "battle" of opposing scientists (Boffey, 1976), provides a good illustration of network effects. Biomedical scientists were particularly likely to oppose nuclear power, while physicists were relatively favorable (Exhibit 13). ANTINUCLEAR is a popular list while PRONUCLEAR is elite, so membership in the NAS differentiates the pros from the cons (Exhibit 12). Also, there are strong network links from the JOHNSON and PROABM lists to PRONUCLEAR (Exhibit 14). Therefore, we would expect NAS members who supported the ABM or the Johnson candidacy, and who are not biomedical scientists, to be relatively pronuclear. The combination of these factors works well in differentiat-

EXHIBIT 15

Percentage of Multiply Listed Scientists (with stated
Attitudes Toward Nuclear Power) Who Favor Nuclear
Plants, as a Function of Three Social Characteristics

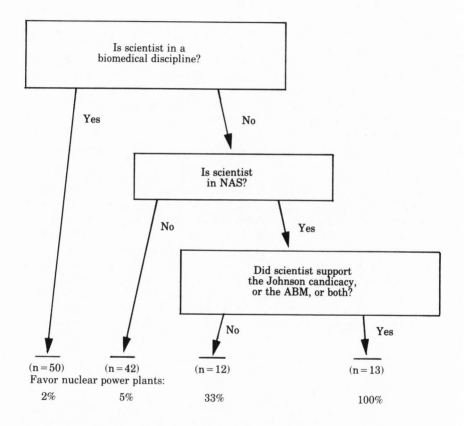

ing scientists who signed pro and antinuclear petitions (Exhibit 15).[6]

It seems to me that the predictability of pro or antinuclear petition signing on the basis of one's scientific discipline, one's status in
the hierarchy of science, and one's positions on prior issues which are
ostensibly unrelated to the nuclear power controversy, is best explained as a consequence of the social networks which exist in the academic science community. The political judgments of scientists are
shaped by their social milieus, just as are those of laymen, and in this

value laden domain, the scientist has no greater claim to wisdom or objectivity than anyone else.

## Notes

1. A name appearing on multiple lists, spelled identically and associated with the same institution and discipline, is easy to identify as a single person. However, misspelling of names, movement from one institution to another, or lack of an institutional or disciplinary label, make it difficult to identify the same person on two lists, particularly when common names are involved. I took a conservative approach, identifying a person as multiply listed only when I was fairly sure that was the case. Therefore, I have probably underestimated the proportion of multiply listed names.

2. Data from petitions were supplemented by biographical references, particularly *American Men and Women of Science*, 13 ed., edited by Jaques Cattell Press (New York: R. R. Bowker Co., 1976); and *Organization and Members 1976-1977* (Washington, D.C.: National Academy of Sciences, 1976). Psychology and history were included in "Social Science." All non-university affiliations were included in the geographic category "other," independent of their actual location. All information was obtained for 88 percent of the singly listed sample and 98 percent of the multiply listed names.

3. To calculate Q between Lists A and B, I crosstab all 367 multiply listed names according to whether they are on A (A) or not (Ā), and whether they are on B (B) or not (B̄), as in the following table where the numbers of names in each cell are designated $a$, $b$, $c$, and $d$. (The number of names common to both lists is a; the number on neither is d.) Then $Q = (ad - bc)/(ad + bc)$. Q is independent of the number of multiply listed names on either list (a + b or a + c). Note that if $b$ or $c$ is small, Q may be large even though $a$ is very small.

|     | A | Ā |
|-----|---|---|
| B   | a | b |
| B̄   | c | d |

4. Ten positive Qs were excluded from Exhibit 14 because they were less than .2 or based on only one name in common between lists. All of these further interconnect elite lists.

5. To further test the apparent isolation between elite and popular blocs, I calculated the expected numbers of multiply listed names which would appear only in elite lists, only in mass lists, and in both types of lists, on the assumption that there was *no* isolation between popular and elite strata.

All multiply listed names were divided into subgroups by the number of lists upon which they appeared, from two to a maximum of five lists. Let $P(p)$ and $P(e)$ be the probabilities of appearing on a popular list or an elite list respectively. For each subgroup, assume that $P(p)/P(e)$ is proportional to: (number of subgroup names on the four popular lists)/(number of subgroup names on the ten elite lists). By definition, $P(e) + P(p) = 1$, so one can calculate these probabilities for each subgroup. In the doubly listed subgroup, the probability of appearing on two elite lists is $P(e)^2$, so the expected number of doubly listed names appearing only on elite lists is $P(e)^2$ times the number of doubly listed names. In similar manner one can calculate, for each subgroup, the expected number of names on only elite lists, on only popular lists, and on both types of lists. The total expected number of names appearing only on elite lists, etc., is the sum of such expected numbers over all subgroups.

Thus, on the assumption of no isolation between the elite and popular lists, only 21 multiply listed names were expected to appear exclusively on elite lists, while the actual number was 54; only 179 names were expected to appear exclusively on popular lists while the actual number was 216. On the other hand, 160 people were expected to cross the elite/popular boundary, appearing on both types of lists, but only 89 actually did ($p < .001$ by chi square test). There is a clear tendency to stay in one stratum, either popular or elite.

6. Kopp (1979) suggests that similar factors predict positions in the debate over fallout from nuclear weapons testing.

# 7
# Growth of Protest

In the last three chapters we have looked closely at the people who participate in technical disputes, their beliefs, and their social links. Now I want to step back and look at the total controversy as my unit of analysis. Each controversy has a life of its own, perhaps spanning decades. Issues and arguments change over time, and different people enter and leave, but the controversy itself persists as an identifiable entity, sometimes growing in strength and sometimes declining. This evolution of protest is the topic of these next two chapters.

Why do some technologies become the focus of widespread public protest while others, no less hazardous, are tolerated? If nuclear power plants are challenged because of the risks to health, and the supersonic transport was opposed for its possibly deleterious effects on the ozone layer of the atmosphere, then why aren't similar protests raised against fossil-fuel power plants which cause an appreciable mortality rate and may change the earth's climate through a buildup of carbon dioxide? Where is the public protest against the automobile, which kills over 40,000 Americans per year and keeps us dependent on foreign oil? Why is there no citizen opposition to the proliferation of extraneous x-rays, drugs, and surgery promoted by the medical and dental industry, while major court battles are waged to stop the Tellico Dam in order to protect a small fish, the snail darter, one of hundreds of species listed, for administrative purposes, as endangered?

These questions are so complex as to be intractable. The most I can attempt is a helpful analysis, breaking the problem into simpler components. We can view the growth of protest movements as a process which takes place in three successive steps, recognizing that the discreteness of these steps exists more in the mind of the analyst than in the real world.

In the first step, a warning against a technology is brought to public attention and is usually noted in the public literature (i.e., the mass media, a book, the *Congressional Record*, etc.). Reasons are given for opposing the technology—beyond simply monetary costs—and some corrective action may be suggested. Responses from the promoters of the technology, refuting the critics, also appear in the public literature at this time.

The warnings may go no further, or it may move to a second step in which a small number of people, organized into at most a few action groups, attempt to stop or modify the technology, perhaps through legal procedures such as a lawsuit, or through lobbying, protest demonstrations, or letter writing campaigns. The essential feature of this second step is that the activity is bounded, or limited, to a small number of well defined groups, and the interest of the mass media is restrained. If the number of protesting units multiplies quickly, and they become joined into a more or less organized network, and many previously uninvolved people are recruited into the protest, then we enter the third step, the regular mass movement so enjoyed by the public media. These steps, and intermediate links between them to be discussed below, are illustrated in Exhibit 16, which provides a convenient structure for the discussion in this chapter. In the next three sections I will describe each step in more detail.

### Step 1. A Warning Is Brought to Public Attention

Many new technologies are initially challenged by the promoters themselves, or their associates, and are settled before we on the outside ever hear of them. Within the community of radiologists and other physicians who treated children's thymus problems, there were some who doubted that these "problems" were serious enough to warrant treatment, but their concerns were kept "inhouse," away from public attention, and the promoters of radiation therapy prevailed.

Inhouse challenges are most likely to occur when there is a preexisting cleavage within the community of professionals involved with the technology. It is unreasonable to expect a proud promoter to be a strenuous critic of his own invention, but if the promoter belongs to one of two competing divisions of a company, or to one of several competing companies, then challenges are likely to come from the competition. In technologies heavily dependent on federal money, government agencies and congressional committees which oversee these funds become integral parts of the involved community, and there is preexisting cleavage along the lines of party or other voting

EXHIBIT 16

Three Steps in the Growth of Protest

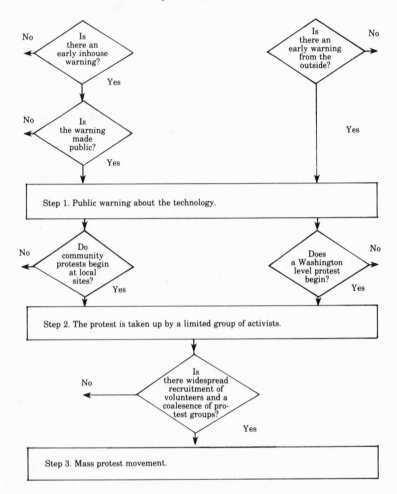

blocs, which encourages challenges to the proposed technology. Several space and weapons systems have been promoted and then stopped by such opposing interests without ever becoming public issues of any import, for example, the nuclear airplane (Lambright, 1967).

The major problem with internally generated criticism is that it is easily overridden. Arthur Kantrowitz recalls that when he was a scientific advisor on the Apollo program there was an internal debate over two alternative ways of getting a payload to the moon. A heavy load could be launched with a single large booster, not yet developed, or the load could be sent up in pieces by multiple launches of small boosters already in production, and then assembled in orbit. Kantrowitz and the majority of advisors favored the small boosters which were relatively cheap and already proven, whereas NASA had committed itself to the development of the huge new Saturn booster. Kantrowitz presented his case to Jerome Weisner, President Kennedy's science advisor, and when Weisner passed it on to the President, Kennedy replied, "This is none of your damn business" (Kantrowitz, 1978; Ames, 1978: 45). Similar instances of presidential suppression of contrary internal technical advice have been documented in the controversies over the ABM and the supersonic transport (Primack and Von Hippel, 1974).

Inhouse challenges rarely reach public attention, though there have been some dramatic instances where insiders have gone public. Physicists Sidney Drell and Marvin Goldberger were two members of a scientific panel convened in 1970 by the Department of Defense to advise on the proposed Safeguard ABM system. Some months later, the research director of DOD testified about the ABM before a Senate subcommittee, claiming that the panel "said that this (ABM) equipment will do the job that the Department of Defense wants to do." Almost immediately Drell and Goldberger contacted the subcommittee claiming that the panel's report said no such thing, and they then went on to severely criticize the ABM, thus bringing their views to public attention (Cahn, 1974; Primack and Von Hippel, 1974).

Within the esoteric world of molecular biology there were methods developed in 1973 which allowed the splicing together of DNA (the genetic material) from different species of organisms. This recombined DNA, when inserted into a bacterium, would replicate as if it were the normal genetic complement, though in fact it could be a completely new hybrid. Some of the early recombinant DNA workers became concerned that these new life forms could be extremely toxic or tumor forming, and that they might escape from normal laboratory confinement, spreading to the population. A clique of these prominent researchers formed themselves into a committee under the auspices of the National Academy of Sciences and published a letter in *Science* requesting a voluntary moratorium on certain recombinant experiments until the hazards could be evaluated and appropriate safeguards designed. *Science* is the largest-circulation scientific journal in the world, and the letter caught the eyes of many people beyond the biology community, triggering one of the major technical controversies of the 1970's (Ames, 1978).

The public may be warned of a technology by an outsider as well as an insider, as when biologist-writer Rachel Carson wrote of the hazards of pesticides in *Silent Spring* (1962), or Ralph Nader attacked the Chevrolet Corvair in *Unsafe at Any Speed* (1965). Isolated writers still serve this function but increasingly, since the 1960's, there has developed a network of scientific, environmental, and consumer groups which have become permanent watchdogs, warning of a range of technologies from smoke detectors and microwave ovens to atomic power and the extraction of fuel from Rocky Mountain oil shale. Some of these organizations have been around for decades, but most came into being since 1960 (Nichols, 1974; McFarland, 1976). The Sierra Club began in 1892 to promote the enjoyment of wildlands and to protect them from spoilage; it played a leading role in the establishment of several national parks. Its membership and power increased rapidly with the growth of the Environmental Movement of the 1960's and its concerns became more technological as it took leading roles in the controversies over pesticides, the supersonic transport, and atomic energy. The Federation of American Scientists began shortly after World War II to lobby for the control of nuclear arms, and though it fell into inactivity soon afterward, it was revitalized in 1969-70 to become a leading opponent of the ABM and has remained active on other science-related issues. Most of the political causes of the remarkable decade of the 1960's had some impact on the changing concerns of these older groups and the formation of numerous new ones, whether it was the desire to apply our scientific skills to the problems of the ghetto, or the revolt against military research during the Vietnam era, the rise of consumerism, or the wreck of the environment by technology. These groups and associated individuals—lawyers, scientists, writers, environmentalists—have a large supporting membership, excellent access to the mass media, and a broad range of publications which they put out themselves. They constitute a potent political force on the Washington scene, in the courts, Congress, and the Executive Branch. Many alumni of this network now serve in prominent government positions, and these personal links provide access to the federal policy process. However, their resources are limited compared to their corporate adversaries, and they seem particularly attracted to fashionable causes, ignoring many that are not.

Sometimes the initial challenge to a technology comes from the grass roots, with opposition arising spontaneously in communities near a site where the technology is to be implemented. The first public opposition to fluoridation was raised in Stevens Point, Wisconsin, led by an elderly man long active in local politics, who didn't want chemicals dumped into the town water supply (McNeil, 1957). The initial opposition to nuclear power plants, back in the 1950's and early 1960's, was raised by local groups over plants proposed for their areas (Mazur, 1975). Several new airports, or proposed enlargements of existing airports, have been opposed by local citizens groups (Milch, 1979; Nelkin, 1974). Recent opposition to very high-voltage electric-transmission lines was generated primarily by farmers whose lands would be crossed by the lines (McGuire, undated; Gerlach, 1978; Tripp, 1978). These local challenges are usually an attempt to protect one's personal interests from infringement by external agencies, particularly the state or large corporations. Initially the technology is seen as unacceptable because it is being sited here rather than somewhere else, though as the protest evolves, the technology may come to be opposed on its own, no matter where it is located.

There are, of course, federal agencies specifically concerned with the identification and assessment of hazards to health and the environment. Early in the century, the Food and Drug Administration used to verify the safety of various substances by feeding them to volunteers, a practice now discontinued, though apparently not for lack of willing participants. This method was not quite as foolhardy as it may appear at first, because the drugs of that time were relatively ineffective, for good or ill. However with rapid strides in pharmacology's ability to do us in, better evaluative techniques were called for.

Since the 1960's there has been a move to increase risk assessment and regulatory activities of the federal government, with the attendant creation of new agencies. Among the most important are the Environmental Protection Agency, the Consumer Product Safety Commission, the Occupational Safety and Health Administration, and the Congress's Office of Technology Assessment. By most accounts, the tasks given these agencies overwhelm the resources which they have available for the work. Also, there has been a reaction in recent years to "over regulation" by government, which is often blamed as a cause of inflation and a detriment to increased industrial productivity, so the future course of government regulation of hazards seems unclear.

There is one more important agent for bringing a warning about a technology to public attention, and that is the intervention of fate. The tragic fire and crash of the Hindenberg Zeppelin marked the end of hydrogen-filled dirigibles. A more recent example is the crash in 1979 of a DC-10 jumbo jet, killing all aboard. Investigation revealed a structural defect, subsequently found in additional DC-10s, as the probable cause of the accident.

Thus, challenges and warnings about certain technologies reach public attention from a variety of sources both inside and outside of the community of promoters, and from the grass roots of the nation as well as Washington. This does not appear to be a particularly efficient screening system since many hazardous technologies are ignored, unless fate steps in, and excessive attention may be focused on a few technologies which are not inordinately risky. However, there can be no doubt that real dangers are discovered and recorded in the public literature for everyone to see, if they would look.

## Step 2. The Warning is Taken Up as a Bounded Protest

Once a warning is made public, it may or may not be taken up by a group that is willing to act as an adversary to the promoters of the technology (Lawless, 1977). Technological products for which warnings have been raised, but public protests were never strenuously pursued, include birth control pills and numerous other drugs, color television, microwave ovens, weather modification, medical and dental x-rays, numerous food additives, aluminum cookware, wood burning stoves, gas furnaces, electric wiring in the home, chlorination of water, swine flu vaccine, several surgical procedures, smoke detectors, hybrid seeds, the internal combustion engine, computer data banks, dams, and so on. In most cases, the warnings were little noticed and barely remembered.

In cases where the warning is taken up, the incipient protest tends to be bounded in the sense that the number of protesters is relatively small, only a few easily identifiable groups are involved, and they are limited to a few geographic sites. Usually the attention of the mass media is limited at this point, and the general public is uninformed, perhaps unaware, of the controversy.

The early protest will be played out either at the local level, in a community where the technology is to be sited, or at the federal level, which usually means in Washington. The particular level, local or federal, is determined in large part by the manner in which the technology is to be deployed. If it is to be deployed in designated sites around the country, as in the cases of fluoridation, nuclear power plants, or electric transmission lines, then the earliest protests occur near proposed sites. If the technology is not associated with particular sites, either because it is spread uniformly throughout the population, or because it is national in character, then the protest will focus on Washington. Thus the controversy over the airbag, which was to be required in all automobiles as a crash safety device, took place wholly in Washington because there were no salient locales as alternatives. Since the ABM was a national-level weapons system, its early protest also focused on Washington, though later, when certain cities were designated as proposed missile sites, these became foci for local disputes.

Community conflicts over nuclear power plants and fluoridation have been studied in detail, and a good deal of interesting descriptive material has been developed, but without much progress in figuring out the complex dynamics of these situations (Crain, et al., 1969; Nelkin, 1971; Lewis, 1972). We cannot predict with much success which communities will accept fluoridation and which will reject it, despite many attacks on the problem. Some low level generalizations are possible, however, because certain patterns of community conflict occur repeatedly. Saunders (1961) has outlined stages which appear in a typical community dispute over fluoridation, and Jopling (1970) has done the same thing for early nuclear power plant siting disputes. These two schemes, compared in Exhibit 17, are similar, though their authors apparently worked independently of one another. I have indicated commonalities in the third column of Exhibit 17, which seem applicable to community protests over electric transmission lines and airports as well.

I am concerned that my emphasis on commonalities across community disputes will obscure the fact of marked differences among them. Each community is idiosyncratic in important ways, and furthermore, there are systematic differences between technologies. For example, communities typically used a referendum to decide whether to fluoridate or not, and the local dispute focused on the campaign, drawing a large portion of the citizenry into the controversy, at least by way of casting a vote. In contrast, the important public decisions for siting a nuclear power plant are made by state and federal licensing agencies so local citizens do not have much say in the matter and few become involved in the licensing hearings. In spite of these important differences, I have chosen to treat community conflicts as if they were similar units because my concern here is with the growth of protest to a larger scale. For this purpose, it is not very important to know what goes on within a given community. What is crucial is

EXHIBIT 17

Stages in a Local Siting Controversy

| Fluoridation (Saunders, 1961) | Nuclear Power Plants (Jopling, 1970) | Common to Fluoridation and Nuclear Power Plants |
|---|---|---|
| A small group of proponents discuss the desirability of local fluoridation. | | Initial plans to introduce the innovation. |
| Proponents gather support; open opposition may appear. Formal proposal to have fluoridation approved is submitted to the appropriate government authority. | The electric utility proceeds to satisfy federal and state regulatory requirements regarding site. First public opposition to, or questioning of, the utility's proposal. | Proponents submit a proposal to appropriate governmental authorities for approval; the first opposition appears. |
| Broadening of public opposition; local activity against proposal; formation of active groups for and against fluoridation; public debate; news organs become involved; campaigning along political lines. | Early opponents warn the public of the dangers inherent in the proposal. Broad public opposition emerges; civic groups form to resist the proposal. The electric utility, reactor industry, and federal agencies are unable to allay public opposition; national publicity results; dispute on technical issues discloses disagreement among experts. Public demonstrations and campaigns against the electric utility and its proposal. | The opposition gains broad public support; formation of active civic groups; public meetings, demonstrations, and election-like campaigns; news organs become involved; disagreement among experts. |
| The governmental authority makes a decision on the proposal. The defeated party either appeals against the decision or raises the issue again. | The electric utility company withdraws its proposal, or a governmental authority (e.g., the judiciary), decides the issue. The defeated party may appeal or prepare for a new confrontation. | A decision is reached by some governmental authority, or else the proposal is withdrawn. But the issue is not necessarily settled. The defeated party may appeal against the decision or raise the issue again at a later date. |

the growing number of communities experiencing disputes, for they serve as building blocks, all of a kind, which become linked together into a national coalition.

## Step 3. Mass Movement

In 1977 farmers in both Minnesota and upstate New York began protests against very high-voltage electric transmission lines which were to run across their lands. These were new types of lines, of higher voltage than had been used previously. Since many farmers resisted initial utility attempts to obtain easements through their properties, the utilities exercised their legal rights to obtain these rights-of-way forcibly. A major issue in the controversy became the utilities' violation of individual rights to land. A second issue was the hazard from electromagnetic fields generated by the lines, a hazard which the protestors claimed to be greater than the utilities recognized. This technical argument was developed by two medical researchers from upstate New York who had conducted some studies on the effects of fields on living organisms. Industry experts totally rejected these claims, setting off an argument which was discussed in Chapter 3. The challenge researchers brought national attention to the lines, if only briefly, by their appearance on the television program, "60 Minutes." Also, their arguments brought some favorable rulings from the New York Public Service Commission, including a requirement for a wider safety corridor along the route of the power line.

There was a good deal of popular support for the farmers' cause in both states, and the power lines became an issue in gubernatorial politics, particulary after some incidents of sabotage against construction work, which were dramatized in a made-for-television movie that was shown nationally. In spite of the efforts of the farmers and their allies, the lines in both states were completed in 1979, and the protests appeared to be over, though it is too soon to be sure that they will not reemerge. Throughout the two-year controversy, from the initial warning to the completion of the lines, the protest remained bounded, being limited to fairly well-defined groups at the two sites; there was not even much communication between the Minnesota protestors and those in New York (McGuire, undated; Brooks, 1978; Gerlach, 1978; Tripp, 1978).

In many ways the fluoridation controversy is similar to the power line controversy. Both began in small communities which had been chosen as early sites for a new technology. In both cases, vocal citizens feared that the technology was hazardous, and perhaps more important than the hazard, they felt that the technology was being forced down their throats in spite of their objections. These activists became the leaders of community revolts. At this point the stories diverge. In two years of transmission line controversy, only two communities protested, partly because there weren't many other sites where very high-voltage lines were being put up. In contrast, several communities considered adopting fluoridation within the first two years of that controversy, so there were many sites available for protest. Furthermore, several of the early protest groups managed to reject fluoridation, while both of the transmission line protests failed. The two transmission line protests remained bounded and virtually isolated from each other. In contrast, the fluoridation protest quickly burst its bounds; the number of community protests increased rapidly, drawing in numerous previously uninvolved people during the course of referendum campaigns. Formal and informal links grew among the local protest groups, lending support and exchanging information. Within a very short while, the opposition to fluoridation had developed into a mass protest movement of national proportions.

Why did the transmission line protest remain bounded and eventually fail, while the fluoridation protest mushroomed? A strategy that has succeeded in other protest movements, not only fluoridation but nuclear power and the ABM as well, has been to promote local protest groups, using them as building blocks to form a national coalition. What would have happened if an organizer had encouraged communication between the Minnesota and New York groups, forming an alliance of mutual support? The challenge medical researchers, who had effectively raised the hazard issue in New York, bringing it to a national television audience, could have lent weight to the Minnesota protest. The two states, now linked together, might have promoted protests in other locales where very high-voltage power lines were being discussed (California would have been a good start) and then brought these groups into the coalition. Perhaps the building momentum would have produced a national protest.

A coalition of local protest groups is one of two pillars of most successful national movements, the other being a strong effort in Washington to influence policy through the federal courts, Congress, and the agencies and departments of the Executive Branch. The fluoridation controversy was unusual in that the opposition worked mostly at the community level, without an effective Washington lobby. In most other cases coalitions have tied the community protest groups to Washington, the usual belief being that skirmishes in the field necessarily have a limited effect, and that the death blow must be struck in the capital at the center of power. The giant nuclear power controversy may be taken as a paradigm of this philosophy. The very first antinuclear protests were held at local sites, and over the years nearly every nuclear power plant in America has encountered some sort of organized local opposition. These site disputes sometimes have caused delays in construction and modifications in plant design, often running up costs a great deal, but few plants have been stopped. In most cases the outcome of licensing hearings has been a foregone conclusion (Ebbin and Kasper, 1974). The resources of the movement have been more effectively spent at the federal level, where the coalition has attempted to cut the umbilical cord which has supported the nuclear industry with all manner of government subsidies and favorable regulation (Metzger, 1972).

The technical protests which grow to the size of a national movement have a number of features in common with large nontechnical movements (Gerlach and Hine, 1970; Obershall, 1973; Kriesberg, 1973), so it is possible to close this section with a fairly robust picture of a generalized movement. With a few variations, the controversies over nuclear power, fluoridation, and the ABM all fit this pattern: Local opposition groups emerge with officers, by-laws, and dues; and these become linked by various personal and organizational ties into a national coalition which supports campaign and lobbying efforts, both locally and in Washington. (Dispersed local protests may precede the Washington effort, as in the nuclear power controversy, or follow the initial Washington action, as in the ABM case.) A few individuals on each side of the controversy emerge as nationally known spokespersons, and opposing, hostile camps become clearly identified, one the establishment side supporting the technology, the other the challenge side made up primarily of voluntary organizations. Experts buttress each side's position with technical arguments which may seem contradictory to the layman. New members are recruited to the challenge side, often from among friends and acquaintances of those already involved. Frequently new members join as a bloc when organizations to which they belong merge with the coalition of protest groups. Public demonstrations help draw the attention of the mass media to the controversy, and the carnival-like atmosphere of these demonstrations may attract protestors whose interests are transient. Movement participants develop a shared outlook which emphasizes the hazards of the technology and their confrontation with the establishment. This outlook is frequently expressed in a stereotyped rhetoric which fits into a pattern of attack and rebuttal with the equally stereotyped rhetoric of the promoters of the technology.

The evolution of a protest into a mass movement is critical for public participation, because it is only at this stage that opportunities are available for lots of people to add their efforts together to produce a desired change. Most of us would have no effect working alone, nor would we receive any social support to sustain our actions. I may write letters to editors and congressmen about any of my favorite problems, but these will be straws in the wind if there are no other letter writers, and if I have no better ways to mobilize opinion and to apply pressure. The average person cannot effectively lobby Congress, or obtain television coverage for a favored cause, or make speeches on a campus tour. But if he cannot start a movement, he can join one that already exists and that provides easy channels for protest, such as letter writing campaigns, marches, meetings, newsletters, rallies, and the like. The options ususaly available to someone who wants to express his concerns are limited: either waste one's efforts on solitary protests with little chance of success, or join a currently-running protest movement and pool resources with other sympathetic souls. Joining a preexisting movement also offers the emotional reward of marching with one's fellows in a just cause, a feeling denied the solitary activist.

## Causes of Protest

I will close this chapter with a question raised at its beginning: Why do some technologies become the focus of public protest while others do not? An answer may be approached (if not reached) from two different viewpoints.

The first viewpoint assumes that the answer resides in some feature of the technology itself: *Technologies with characteristics a, b, and c are more likely to be opposed than those without such characteristics.* What then are those characterisitcs? It seems to me that there are a few which are crudely plausible, though none without its problems. The characteristic which seems to work best is newness. Technological innovations are more likely to be opposed than technologies which have been around a long time. Of course, the vast majority of technological innovations are not opposed by the public. Furthermore, the important case of nuclear power is problematic. Enrico Fermi and Leo Szilard obtained a patent for a nuclear power plant during the

Second World War, and commercial demonstration reactors were developed in the mid-1950's when nuclear powered submarines were already in service. The very first controversy over a nuclear power plant occurred in Detroit in 1956, and a few more sites were contested in the early 1960's, but for practical purposes, the nuclear power controversy really emerged in the late 1960's, reaching massive proportions after the mid-1970's, which is 20 to 30 years after the innovation was introduced, depending on where one dates the introduction. Can we really consider nuclear power to be new by now? The other side of the coin is that most people are not accustomed to nuclear power plants even now, and in that experiential sense they may still be regarded as innovations. However this is resolved, if we leave nuclear power aside and scan the range of technologies which have been protested, a large proportion do seem to be new developments rather than established items. Furthermore, some of the particularly hazardous technologies which have largely escaped protest are those which have been around for a long time, for example, the automobile, and medical and dental x-rays. Thus I suggest as a very low-level generalization that innovations are more likely to be opposed than established technologies.

A second characteristic which seems important, particularly in cases where protest comes from the grass roots, is that the technology or its hazard is imposed on the local citizenry in a compulsory manner rather than allowing voluntary acceptance or rejection by each individual. To some extent the local challenges to fluoridation, electric transmission lines, and airports are attempts to protect one's own interest from infringement by external agencies, particularly the state or large corporations. There had been no public opposition to automobile seat belts until legislative attempts to make their use compulsory, at which time protests began to appear, for example:

Statistics cite the . . . lives possibly saved by the use of seat belts, but they fail to enumerate the number of deaths caused by people being burned to death or otherwise fatally trapped in a vehicle because they were strapped in . . . Do we as Americans want to retain the right to determine whether or not we wish to die strapped in a machine, . . . or do we still wish to retain the freedom of decision which we have inherited as a part of our American birthright? (Nolan, 1973).

In addition to considerations of newness and compulsory adoption, it seems that some sort of balancing of the benefits and risks of a technology affects the likelihood of protest. Thus, as many problems as the automobile causes, we tolerate it because its benefits are seen to outweigh its costs. The problem here is that it is usually very difficult to obtain objective measures of the benefits and risks of a technology, particularly a new one for which there is little accumulated experience. Even when there is appreciable experience, as in the case of the automobile, it is difficult to account for all the diverse benefits and costs, and it is nearly impossible to convert them into the same units (e.g., the dollar equivalent of a life lost) so that we can determine if the amount of benefit is greater than the amount of the cost. Furthermore, when there is a dispute over the technology, we know that its proponents tend to overestimate benefits and underestimate costs, while its opponents move in the opposite direction (Chapter 5). Thus, to some extent, one's position toward the technology—for or against—determines one's estimate of its net worth, rather than the other way around.

I am left with the belief that characteristics of the technology do have a small effect in influencing whether or not it will be opposed, but much of the answer must be found in a different approach to the problem. Throughout this chapter I have pictured the growth of protest as a process which can wax or wane depending on all sorts of factors along the way. A warning may or may not come to public attention. It may or may not be handled adequately by government action. Groups may or may not have the resources and inclination to take up the protest, and if they do, they may be more or less successful depending on their strategy, the response of the mass media, the degree of public support, etc. From this viewpoint, the growth of protest is seen as a historical process which must be understood within the context of the particular time and place where it occurs. I will pursue this viewpoint in the following chapter, examining fluctuations in the rise and fall of controversy and their relationship to specific interests which occupy the nation at various times.

# 8
# Rise and Fall of Controversy

*(with Peter Leahy)*

The rise and fall of various antitechnology movements is determined, to a great extent, by rising and falling levels of public interest in larger issues which are relevant to the technologies. For example, the movement against nuclear power gained great strength during the late 1960's when the American public became massively concerned with the issues of environment and pollution. It is during such periods that protest leaders are best able to raise funds, recruit members, find allied organizations, and have access to the mass media (Zald and Ash, 1966). In short, these are the times when social resources become widely available to support movement activities.

My concern in this chapter is limited to protests which have become mass movements, reaching Step 3, as defined in the preceding chapter. The focus will be on the familiar movements for which the best data are available: those against nuclear power plants, fluoridation, and the ABM. By developing quantitative indicators of movement activity, I hope to show that fluctuations of opposition to the technologies follow an orderly sequence in each controversy. First, challenges from activists increase during periods of national concern with a major issue closely related to the controversy. This national concern provides a supportive milieu in which the activists can obtain new members, money, media attention, etc. As activism increases, spurred on by the new availability of societal resources, the attention of the mass media turns increasingly to the protest, partly to report flects concern with some larger national issue such as the degrada-

This chapter includes excerpts from an article by the author, entitled "Opposition to Technological Innovation," which originally appeared in *Minerva*, XIII, 1, Spring 1975, pp. 58-81; and from an article by the author and Peter Leahy, entitled "The Rise and Fall of Public Opposition in Specific Social Movements," which originally appeared in *Social Studies of Science*, Vol. 10, 1980, pp. 259-84. Used with permission.

the activities of the protestors and partly to carry protechnology propaganda which is designed to refute the challengers. The nation then becomes a spectator to the dispute through the press and television, and the increasing prominence of the controversy in the media is followed by increasing opposition to the technology within the wider public. Mass media attention soon drops off, perhaps because of a saturation effect, or more likely because of competing issues vying for attention, and opposition among the wider public drops off as a result. If the activists raise their level of protest again, the mass media again increases its coverage, and opposition again increases among the wider public.

This model is summarized by three hypotheses:

Hypothesis 1. The greater the *national concern over a major issue* that is complementary to a particular protest movement, the more easily resources can be mobilized for the movement, and therefore the greater the *activity of protesters*.

Hypothesis 2. As the *activity of protesters* increases, *mass media coverage* of the controversy increases.

Hypothesis 3. As *mass media coverage* of the controversy increases, the *wider public's opposition* to the technology increases.

The model depends heavily on the distinction between activists and the general public. The activists described in Chapter 3 expressed their opposition to a technology like fluoridation in terms of larger national issues like socialism and individual rights. During periods of high national concern with these larger issues, the activists are spurred, aided by sympathetic supporters. The wider public is essentially passive, but as they become aware of the controversy through the media, they become increasingly skeptical of the technology, increasing their opposition responses in opinion polls and referenda.

## Quantitative Indicators

It is possible to point out periods of intense activity and periods of calm in the history of each controversy, but these judgments are often crude, so it is useful to chart yearly fluctuations with quantitative indicators. One such indicator is the coverage that mass media periodicals give to the technology and to the dispute around it, as measured by the yearly number of articles indexed under each technology in *Readers' Guide to Periodical Literature*. The *Readers' Guide* has an obvious shortcoming, ignoring television and the daily newspapers, but it does give a good view of the interests of a wide range of magazines and journals in the United States.

The opinion of the general public toward a technology may be measured by repeated opinion polls. It is rare to find the ideal situation of the same polling organization asking the same question, frequently, at fixed time intervals, using samples drawn in a similar manner each time. However, usable time series are available for nuclear power plants, fluoridation, and the ABM.

The activity of protesters is especially difficult to measure. It is desirable to have an indicator that would be applicable across all controversies, but none is available. For the early years of the nuclear power controversy, it is reasonable to count the number of nuclear power plants, or proposed plant sites, where citizen groups had recently intervened as a measure of activism. However by the early 1970's nearly all such sites were opposed, so the indicator becomes saturated, unable to measure further increases in protest. Also, important aspects of protest, such as lobbying in Washington, may occur independent of activity at local sites, particularly as the controversy has matured. The yearly number of defeats for fluoridation in community referenda is a useful measure of leadership activity since the organization of a referendum usually indicates organized opposition, particularly when the proposal is defeated (Crain, et al., 1969). Since so many communities were available as fluoridation sites, this indicator does not have the saturation problem of the nuclear plant sites; also, the bulk of antifluoridation activity went on at the community level with little Washington activity, unlike the nuclear power case. No adequate activism indicator is apparent in the ABM case; the omission is not serious in this short controversy, however, since observers agree when the one period of intense protest activity occurred.

The predominantly parallel indicator trends for each technology, shown in Exhibits 18 through 20, indicate that these indicators provide a reliable picture of periods of generally rising and falling controversy. (There are some deviations from the parallelism which shall be discussed shortly.) It is sometimes claimed that movements follow a natural history of rise, peak, and decline as we see in the case of the ABM. However, the movements against nuclear power and fluoridation show multiple peaks of opposition, so the single peak notion is obviously oversimplified, and fluctuations in opposition remain to be explained.

**Protest Activity to Media Coverage to Public Opposition**

It is convenient to postpone consideration of hypothesis 1, so this discussion will begin with hypotheses 2 and 3. That is, as protest activity increases, there is a corresponding increase in mass media coverage of the controversy; and that as mass media coverage increases, there is an increase in opposition to the technology among the wider public. We will examine each controversy to determine the extent to which this general picture is correct.

## ABM

In September 1967, President Johnson decided to deploy the Sentinal ABM system to protect American cities from a Chinese attack, which would be easier to defend against than a sophisticated Soviet attack (Adams, 1971; Halsted, 1971). There had already been low levels of opposition to ABM from scientific advisors and government decision makers, but opposition increased markedly after the President's deployment decision, particularly in the Senate and at proposed ABM sites in Seattle, Chicago, and elsewhere (Cahn, 1974). In March 1969, President Nixon dropped the Sentinal system, substituting the Safeguard ABM which would defend American missile silos rather than cities, and which was perceived as a defense against Russian as well as Chinese attack. Over the next few months opposing blocks solidified; the Nixon administration strongly backed Safeguard. In August, the Senate barely defeated an attempt to halt development. This was the high water mark of the opposition, and subsequent attempts to stop the ABM had less support and intensity. By 1971, public involvement in the ABM was over. In 1972, the United States and the Soviet Union agreed to limit ABM development to two sites in each country. This was the shortest of the three controversies considered here and the only one that is clearly over at this writing in 1981.

There is no adequate indicator of leadership activity in the ABM case, but the omission is not serious in this short controversy since there is consensus among observers that the one period of intense protest activity was 1968-69. At the beginning of 1969, after this activity was well under way, there was a sharp rise in the number of national periodical articles on the ABM and on the debate which was forming around it. Three national opinion polls were taken on the ABM by the

EXHIBIT 18

The ABM Controversy

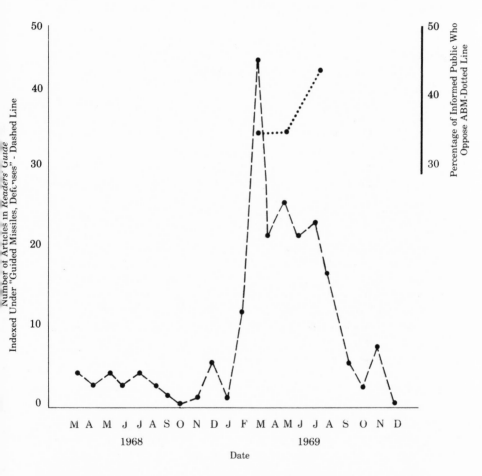

Gallup organization, all in 1969, and these show a rise in opposition among the informed public[1] which was observable about six months after the peak of mass media coverage. This time lag is not precise because of the crudeness of my indicators and the unknown distribution of television and newspaper coverage of the controversy, but the data on hand support the hypothesized sequence of rising activist protest, rising media coverage, and rising opposition among the wider public (Exhibit 18).

## Fluoridation

The first communities to fluoridate encountered little opposition, but then a raucous conflict developed at Stevens Point, Wisconsin, an area of concerted activity by dentist-proponents, and fluoridation was defeated by town referendum in 1950. Since that time fluoridation has spread though each year some communities still reject it in referenda. I have taken yearly number of referendum defeats as my measure of activity among antifluoridation leaders, but this indicator has a built-in time lag since a year or more may pass between the initiation of a referendum and the vote. Comparison of trends in the early years of the controversy is further complicated because the *Readers' Guide* was published biannually until 1965, so the dating of early peaks of mass media coverage is relatively imprecise. Several local groups of activists began referendum campaigns shortly after the Stevens Point defeat of 1950, and many defeated fluoridation by 1952. These disputes are discussed in the periodical output which peaks in about 1952 (Exhibit 19). Public opinion polls show increasing opposition between ˉ952 and 1953, following the article peak, and then decline. We cannot follow opinion change between 1956 and 1959 because the question changed.[2] There is a sharp rise in referendum defeats in 1964 followed, in 1965, by a new peak in mass media coverage and also by a relative high in opposition in the opinion polls. All indicators again decline until 1970 when there is another rise in referendum defeats accompanied by a rise in periodical articles. Unfortunately no opinion poll was taken in 1971, but the 1972 poll shows an increase in opposition from 1969.

## Nuclear Power Plants

The first citizen intervention, against Detroit's Enrico Fermi nuclear power plant, occurred in 1956 (Alexanderson and Wagner, 197ᵒ), but is isolated in time. There was a great deal of mass media coverage of atomic power at that time, almost all of it positive, particularly the discussions of President Eisenhower's "Atoms for Peace" program which encouraged the development of civilian nuclear reactors. The opponents were barely noticed.

The quantitative indicators of controversy in Exhibit 20 begin in 1960, shortly before a small cluster of plant interventions in the early 1960's which did attract mass media attention, particularly the dis-

EXHIBIT 19

The Fluoridation Controversy

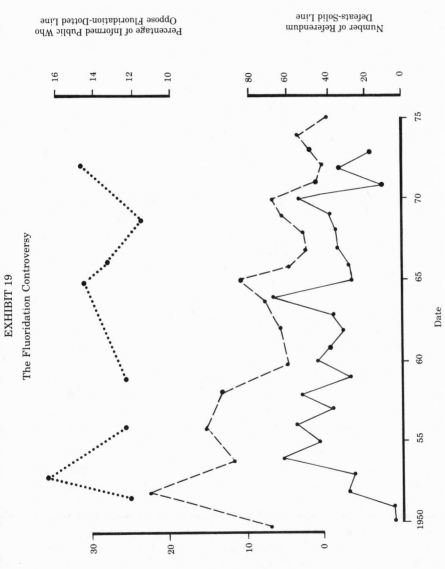

Number of Referendum
Defeats-Solid Line

Percentage of Informed Public Who
Oppose Fluoridation-Dotted Line

Number of Articles in *Readers' Guide*
Indexed Under "Fluorine Content" or
"Water Supply - Fluoridation"-Dashed Line

pute at Bodega, California (Novick, 1969). No public opinion trend data are available prior to 1964, so it is impossible to determine if there was a rise in opposition in response to the 1964 peak of mass media coverage, as hypothesized. We can see, however, that opposition in the polls diminished as the output of periodical articles waned. There was little protest in the mid-1960's, but in 1968 local groups of activists began to intervene against most nuclear power plants that had been proposed. Periodical articles increased to cover these disputes and to carry propaganda put out by both sides. Opinion polls, now appearing yearly, show a peak of opposition among the wider public in 1970 following the peak of mass media coverage in 1969. It was difficult to tally the numerous plant interventions after 1970, but few plants were unopposed in 1971. It appears to me that the activity of protest leaders declined in 1972, but objective indicators are not available to verify that impression. In any case, there was a decline in mass media coverage, and in opposition on opinion polls through 1973.

The antinuclear activists showed new strength after 1973 (Lapp, 1975; Alpern, Bishop, and Cook, 1975; McFarland, 1976). This is reflected in rising mass media coverage throughout the period 1974-76, with a great deal of attention paid to a California referendum in 1976 which attempted to impose a moratorium on nuclear power but failed. Opposition in the opinion polls rose from 1973 to 1974, and from 1975 to 1976, an apparent response to the rising mass media coverage of that period. Unfortunately we cannot say what happened between 1974 and 1975 because of changes in questionnaire wording and polling organization in 1975.[3] After 1976 both media coverage and negative opinion fell off, then rose slightly again in 1978 and then rose massively in 1979 following the highly publicized accident at Three Mile Island, when protest activity reached its highest level to date (Kasperson, et al., 1980).

The trend data presented here are crude, erratic, and incomplete. To see in them support for my hypotheses requires a sympathetic eye, and others may reasonably disagree because of the large errors of measurement. The accident at Three Mile Island allows a finer test of the hypothesized link between media coverage and public opinion because, in the year following the accident, the Harris Poll took numerous opinion surveys, closely spaced in time (Mitchell, 1980). This fine-grained opinion trend can be compared to weekly fluctuations in coverage of the accident on television network news, in *The New York Times,* and in the major news magazines, as shown in Exhibit 21.[4] In

EXHIBIT 20

The Nuclear Power Plant Controversy

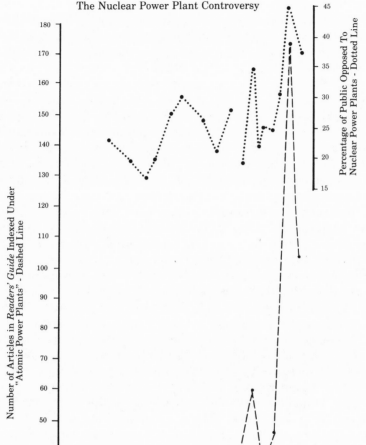

Date

order to remove erratic fluctuations, the media trends have been smoothed by the method of running medians (Tukey, 1977).

The accident began on March 28, 1979, and in the week following, almost 40 percent of television-network evening news was devoted to it. Coverage in the news magazines was necessarily delayed by one week, but both *Time* and *Newsweek* ran cover stories in April. By June the story had disappeared from the news magazines and appeared in only occasional short pieces on television until about October, when there was a second, much smaller, rise in coverage to report the final work of the Kemeny Commission which had been appointed by President Carter to investigate the accident. The Commission's report was released at the end of October, but the media had been anticipating it, reporting related events throughout October.

Fluctuations in public opinion throughout 1979 are precisely what would be predicted from the coverage-opinion hypothesis. The proportion of the public opposing the building of more nuclear power plants rose sharply after the first burst of coverage. This is hardly surprising since the specter of the accident would be expected to increase opposition regardless of any independent effect of the quantity of coverage. However, one would not expect, on this basis alone, that support for nuclear power would rebound within two months, as soon as the media coverage had fallen away, yet that is what happened, in accord with the coverage-opinion hypothesis. Furthermore, a clear, short-term increase in public opposition appeared again during October and November (with rebound by December), coinciding perfectly with the secondary peak of media coverage at the time of the Kemeny Commission's final work. This is completely in accord with the coverage-opinion hypothesis, and is not otherwise explainable in any obvious way.

Why does public opposition increase as media coverage increases, even in cases such as fluoridation where the media do not carry an obvious bias against the technology? Perhaps the prominence given to disputes between technical experts over the risks of the technology makes it appear dangerous to the public. Persons experimentally exposed to both positive and negative arguments about fluoridation were more likely to oppose it than persons who had not seen any of the arguments at all (Mueller, 1968). Fluoridation is more likely to be defeated in a referendum when there has been heated debate than when the campaign has been relatively quiet (Crain, et al., 1969). During a local controversy over a proposed nuclear waste storage facility, residents who had heard about the controversy were more

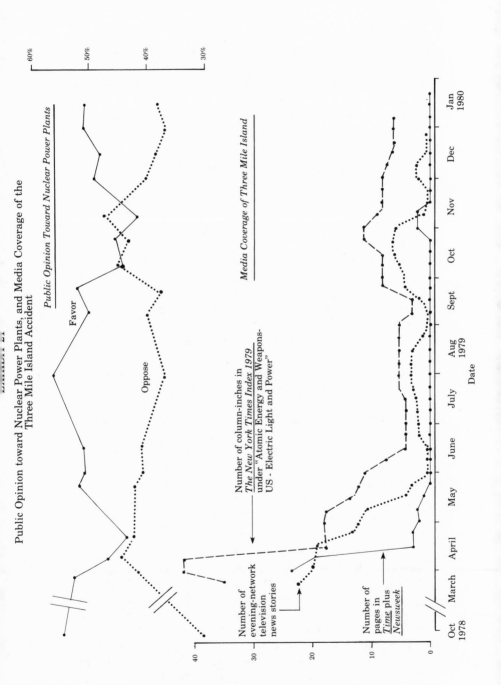

EXHIBIT 21

Public Opinion toward Nuclear Power Plants, and Media Coverage of the Three Mile Island Accident

likely to oppose the facility, and to consider it unsafe, than residents who had not heard about the controversy (Mazur and Conant, 1978). If doubt is raised about the safety of the technology, many in the public prefer to err on the side of safety, as if saying, "When in doubt, reject the technology—better safe than sorry" (Mueller, 1966; Sapolsky, 1968). Thus, the appearance of dispute works to the benefit of the opponents of the technology.

## What Drives the Controversy?

We have seen that fluctuations in a controversy can be represented by more or less concomitant changes in protest activity, public opinion, and mass media coverage. We can go farther now and ask what drives the controversy? Why does it increase in intensity at some times and decrease at others? Three explanations will be evaluated here. First, it is sometimes claimed that the activity of the opposition is a reaction to the activity of proponents. Second, controversy may increase in response to events which make the technology look particularly hazardous. Third, fluctuations in the controversy may be tied to exogenous fluctuations in national concern with larger related issues.

Do the activities of protesters increase as a reaction to increased activity by proponents? The increased activity of nuclear power opponents after 1973 did coincide with nuclear proponents' enthusiastic response to the "energy crisis." On the other hand, nuclear opposition fell off in the mid-1960's, though reactor orders from electric utility companies increased rapidly during that period, and opposition reached its highest level to date in the late 1970's after new reactor orders had dropped to almost nil. Active opposition to the ABM fell off after the climatic Senate vote of 1969, even while the Nixon administration proposed enlargement of the ABM program. Thus, while some opposition activity is probably a reaction to promotional activity, it is easy to find instances where this had not occurred.

Does the level of opposition increase in response to events which make the technology look particularly hazardous? Anyone who has followed the slow public response to the documentation of smoking hazards may be skeptical, though more dramatic events might have a greater impact. Nuclear reactor accidents which critics viewed as serious occurred and were reported in the media in 1952 (Chalk River, Canada), 1957 (Windscale, England), 1961 (Idaho Falls), 1966 (Fermi), 1975 (Browns Ferry), and 1979 (Three Mile Island) (Weinberg, 1979).[5] These dates bear no *overall* relationship to the fluctuations in opposition to nuclear power, as they appear in Exhibit 20. However, the Browns Ferry accident of 1975 does coincide with a period of rising controversy, and there can be no doubt that the immediate public reaction to Three Mile Island was an emphatic rise in opposition, including a demonstration in Washington only two months afterward with over 70,000 protesters, and another in New York four months after the accident with 200,000 protesters.

Was the public reaction a direct result of the severity of the accident, with the media serving simply (passively) to present the facts, or did the public react more to the quantity of media coverage than to the severity of the accident? Within the controversies examined here, heightened mass media coverage regularly brings about increased opposition among the wider public. Past nuclear accidents have not regularly brought about increased public opposition to nuclear power. Faced with a new situation involving both an accident and heightened coverage, it seems reasonable to suggest that it was the quantity of coverage that caused the public reaction and not the accident per se.

The occurrence of accidents—events which emphasize the hazards of a technology—does not appear to drive the controversy, though perhaps an event as major as Three Mile Island is an exception. I have already argued that the opposition is not to any great degree a reaction against the activities of the technology's promoters. The last explanation to be considered here is that fluctuations in the controversy are driven by exogenous fluctuations in national concern with larger issues related to the technology.

We have seen that protest activists express their opposition to a technology in the ideological context of broad national issues such as the environment, or the rights of the individual against the state. Perhaps, then, the rise and fall of activists' protest actions are linked to the rise and fall of national concern with these broader issues. This, of course, is Hypothesis 1, stated at the beginning of this chapter. If this hypothesis is correct, it could explain much of the drive behind a controversy, for as protest activity increases, media coverage of the controversy increases; and as media coverage increases, opposition to the technology increases among the wider public.

## Large National Issues

If opposition to a technology such as the nuclear power plant reflects concern with some larger national issue such as the degradatios of the environment, then a period of rising national interest in that larger issue might lead to a rise in opposition to the particular technology. The larger issue may itself have the form of a mass movement, for example the Environmental Movement, and then we would have the fusion of two movements, one large and one small, each supporting the other in publicity, enthusiasm, membership, fundraising, organization and communication, and strategy.

Three large issues appear to account for the major fluctuations in opposition to nuclear power plants. These are the atomic bomb-testing and fallout concerns of the early 1960's, the pollution and environmental concerns of 1969-72, and the "energy crisis" of 1973 and beyond.

In order to provide an indication of periods of national concern with these issues and others which will be considered shortly, it is convenient to use responses to the Gallup poll question, which has been asked in most years since 1950, "What is the most important problem facing the country today?" The proportion of the population listing a particular problem in a given year is taken as a measure of national concern with that problem.[6] These trends appear in Exhibit 22 where issues pertinent to each of the three controversies are grouped together.

Examining the three issues related to nuclear power plants (Exhibit 22), we see that the periods of major concern coincide roughly with the periods of rising activity in the nuclear power controversy (Exhibit 20). Many of the scientists who warned of atomic weapons fallout were political liberals and thus congenial to the liberalism of the opponents of power plants; several have since become critics of nuclear power plants (Kopp, 1979). A major issue of the early dispute over the proposed nuclear power plant at Bodega, California—second only to concern that an earthquake might split the reactor—was the fear of radioactive fallout from the plant. In a publicity stunt in 1963, opponents at Bodega released 1,000 balloons marked strontium-90 and iodine-131—isotopes found in fallout from nuclear weapons (Jopling, 1970). After the United States and the Soviet Union signed the nuclear test-ban treaty in 1963, national interest in fallout diminished and, with this removal of support, the incipient opposition to nuclear power plants faded.

The resurgence of opposition in the late 1960's coincides with the rise of massive public concern with the environment. Major environmental organizations such as the Sierra Club and Friends of the Earth became deeply involved in the opposition. Many antinuclear

EXHIBIT 22

What is the Most Important Problem Facing the Country Today?

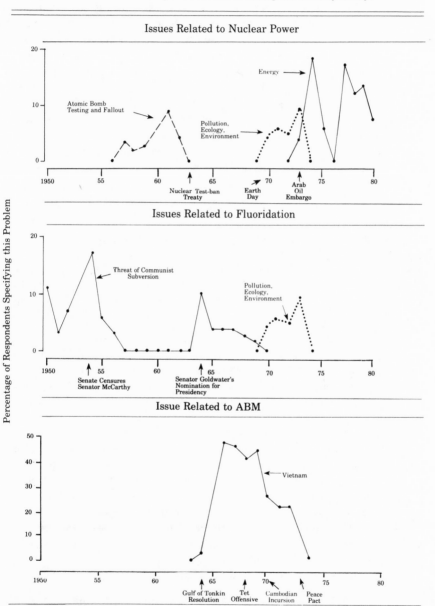

Source: Gallup Opinion Poll

activists entered the nuclear power controversy through their participation in the Environmental Movement.

By 1973 the Environmental Movement was waning. However the Arab oil embargo of 1973 brought on the "energy crisis" as an intense public issue which was clearly relevant to nuclear power. That this would support the antinuclear movement seems, at first, paradoxical since a pressing need for energy might be thought to quell opposition to an energy-producing technology. Instead the increased national concern with energy, like earlier concerns with nuclear fallout and the environment, gave new focus and importance, and a renewed audience, to the movement against nuclear power plants. Antinuclear leaders began to express their opposition in the ideological context of energy conservation and the need for new, safer forms of energy such as solar power.

We may visualize the nuclear power controversy as a surfer riding successive waves which are large national issues—first the nuclear fallout wave, then the environmental wave, and then the energy crisis wave. As each wave diminishes, the technical controversy falls unless it can catch another wave. Each wave, to be suitable, must have clear relevance to the technical issue, and it must be politically compatible, or at least not incompatible, with the leadership of the smaller movement.[7]

As the antinuclear movement changed from one wave to another, the specific issue content of the controversy also changed to become more compatible with the current large issue of national concern. Yearly opinion polls, taken since 1966, have asked people who oppose nuclear power plants for their reasons. Issues of water and thermal pollution and low-level radiation became prominent in the late 1960's with the rise of the Environmental Movement and then passed away with it in the early 1970's (Mazur, 1975). The rapid decline of these particular issues is usually explained as the result of the stringent revision by the Atomic Energy Commission of its radiation standards in 1971, and increased use of cooling towers and ponds to alleviate thermal problems (Nelkin, 1971). I suggest that these issues would have receded at this time irrespective of these policy changes. More recent concerns about moving toward a low energy-consumption society, and developing solar and renewable energy sources (Lovins, et al., 1977), reflect public attention to the national energy problem.

Two large issues appear to account for the three major peaks of opposition to fluoridation. Americans feared socialism and communism in the United States, perceiving in them a threat of pervasive central government which would undermine individual liberties. These concerns were intimately connected with the Cold War, and they reached their height in the early 1950's. They reappeared in the mid-1960's and were associated with the presidential campaign of Senator Goldwater. The other national issue was the environment, including concerns over trace "poisons" such as DDT, mercury, and fluorides. Periods of rising opposition to fluoridation—in the early 1950's, 1964-65, and the late 1960's—coincide with periods of national concern, in Gallup polls, over these two large issues. Environmentalism was not a partisan issue and therefore was politically compatible with the conservative antifluoridation movement as well as the liberal antinuclear movement. Concern over the dangers of socialism could be compatible only with a conservative movement.

I have traced the rise and fall of specific issues in the fluoridation controversy through the editorial content of *National Fluoridation News,* the major antifluoridation newspaper. In the mid-1960's, the period of Goldwater conservatism, the newspaper's major concern was that mass fluoridation violates individual freedom of choice; fluoride as poison was a minor concern. During the period of the Environmental Movement, the newspaper's major issue did become fluoride as poison or pollutant, and its comments on individual rights were minimal (Mazur, 1975).

The ABM debate occurred at a time of intense concern over Vietnam, riots, civil rights, and poverty. Liberal dissent had been growing since the early 1960's; public concern over Vietnam became most intense in the period 1966-69. These issues were acute in the university community; numerous campuses participated in the March 4, 1969, strike against defense-related research which emphasized opposition to the ABM (Nelkin, 1972). Of all these interlocking national concerns, the primary one was the Vietnam War, and this was the wave that carried the ABM into massive controversy. A poster of the time read: "From the people who brought you Vietnam: The Antiballistic missile system." Anti-ABM scientists usually came from the universities where sentiment against the Vietnam War was strong; pro-ABM scientists came predominantly from the military-industrial complex (Cahn, 1974). Anti-ABM senators were usually "doves" while those for the ABM tended to be "hawks." By 1970, the public's concern with Vietnam, as measured on the Gallup poll, was little more than half of what it had been in 1968. While the Nixon administration was pressing ahead with the ABM, the motive power behind the opposition movement had dropped away and the controversy was over.[8]

## Prediction

It must be clear to the reader that the data base used here is grossly inadequate. The *Readers' Guide* index of mass media coverage totally ignores television and the newspapers, makes no distinction between articles slanted for or against the technology, and allows only the crudest indication of the timing of media exposure. Public opinion polls are erratic in timing, question wording, and even sample design. The measurement of protest activity is the most difficult of all without special resources being alloted to that task. Nonetheless, I believe that the data are sufficient to indicate the benefits of this approach.

If the model of fluctuating opposition described here is dependable, and one can identify the major national issues which drive a controversy, then predictions of future controversy activity ought to be feasible. When this theory was first formulated, a prediction was offered that the energy crisis, a newly emerged national issue, would reinforce the opposition to nuclear power plants (Mazur, 1975: footnote 51). This prediction was considered counterintuitive by many commentators at that time who thought that a pressing need for energy ought to alleviate opposition to an energy-producing technology, but it is now clear that the prediction was correct.

There is great deal utility in the prediction/testing approach to social science, so I will make another prediction now, to facilitate the testing of these ideas, recognizing that this attempt may date this book very quickly.

It is 1980, the year after the Three Mile Island accident which some observers think will spell the end for civilian nuclear power in the United States. Should there be similar accidents in the near future, they are probably right. Barring that, the persistence of antinuclear protest in the United States depends on the continued strength of national concern with the energy issue, which can be easily followed by the Gallup poll question about "most important issue." Public concern is quite separate from the underlying reality of the energy issue, so apart from the validity of our energy problems, it is certainly possible that the public will lose interest, as it did in 1976 (Exhibit 22), as long as there are no gas lines or heating oil shortages. A few years of freedom from energy worries would remove the driving force from the nuclear power controversy, whereas the persistence or increase in these concerns would increase protest. There is also the possibility that the energy concern will fall away, only to be replaced by another large national issue which would give new force to the controversy. It is impossible to foresee whether or not that will happen, but if it does, then the new issue will appear at that time on the Gallup poll, it must have obvious relevance to nuclear power, and it will have to be an issue of the political left in order to be compatible with the left orientation of the antinuclear movement.

## Notes

1. ABM public opinion was measured in three Gallup polls available from the Roper Public Opinion Research Center, Yale University. These were taken in March (AIPO 777), May (AIPO 780), and July (AIPO 784) of 1969. Respondents were asked if they had heard about the ABM program, and if so, did they have an opinion about the ABM program submitted to Congress by President Nixon? Those who said "yes" were asked, "Do you favor or oppose the ABM program submitted by Nixon?"

2. Fluoridation opinion was measured in three Gallup polls taken in 1952, 1953, and 1956 (AIPO 0484KO, 0514, and 0559, respectively), available from the Roper Center at Yale; and in five NORC polls taken in 1959, 1965, 1966, 1968, and 1972 (NORC 423A, SRS868, SRS889A, 4050, and 4139, respectively), available from NORC, Chicago, Ill. Respondents in the Gallup polls were asked, "Would you favor or oppose a plan for putting fluoride in the water supplied in this community?" The 1959 NORC poll asked, "How do you feel about fluoridating public water supplies—would you say you're very favorable, favorable, unfavorable, or very unfavorable?" The rest of the NORC polls (1965-72) asked, "What is your opinion on fluoridating public water supplies? Do you feel it is very desirable, desirable, undesirable, or very undesirable?" Since the two NORC questions are similar, they are incorporated into one trend line. The Gallup question is substantively different, focusing on the respondent's own community, and probably elicits more negative responses than does the NORC question. The break in the opinion trend line of Exhibit 19 emphasizes that the Gallup and NORC questions are not equivalent. In all polls, the above questions were asked only of respondents who had heard about the fluoridation of water supplies.

3. Nuclear industry polls were taken in 1964 and 1966-74 under the direction of Underwood, Jordan Associates of New York. The specific question for 1966-72 was, "If someone proposed to build a large atomic power plant here in (name of local area), would you favor or oppose such a plant?" Slightly different questions, still focused on the local area, were asked in 1964, 1973, and 1974. Louis Harris began polling in 1975, asking "In general, do you favor or oppose the building of more nuclear power plants in the United States?" Exhibit 20 uses the Underwood, Jordan series for the period 1964-74 and the Harris series for the period 1975-79. These are not comparable because questions which focus on the local community elicit more opposition to nuclear power than the general question. Cambridge Reports also has

a series covering the period 1975-79, asking "Do you generally favor or oppose the construction of more nuclear power plants?" This series parallels the Harris series after 1976 but is erratic in an unexplained way during 1975, being inconsistent with both the Harris data and the mass media trend line in Exhibit 20. Diverse attitude polls about nuclear power are summarized in Mitchell (1979) and Maynard, et al. (1976).

4. The weekly number of television-network evening news stories devoted to Three Mile Island was obtained from *Television News: Index and Abstracts*. Coverage in the *Times* was measured by the column-inches in *The New York Times Index 1979* under "Atomic Energy and Weapons-US-Electric Light and Power," which was mainly about Three Mile Island. Coverage in the news magazines was measured by the number of pages on Three Mile Island in *Time* and *Newsweek*.

Three Mile Island was a stupendous media event, far surpassing the coverage given two months later to the nation's worst airline accident, the crash of a DC-10 killing 274 people as the result of a faulty engine mount assembly. The power plant accident occurred at a time of public concern with nuclear power, so media people were sensitized to the issue and easily attracted to the event. By a quirk of timing, the antinuclear movie *The China Syndrome* was released to theaters across the nation just days before the accident. The event itself had great human interest—the week-long struggle with the bubble, the heroics of the nuclear engineers in averting disaster, the entry of the President, the exit of the frightened populace—all this had the drama of a soap opera, to be followed day after day. Also, the plant site in Pennsylvania was easily accessible to reporters in nearby New York and Washington (Sandman and Paden, 1979). Each of these factors encouraged high coverage.

5. Apparently a very serious accident occurred in the USSR in about 1973, but this was not publicly known at the time.

6. The number of polls asking this question in a given year ranges from zero to four. In years with multiple polls, averages were taken. Usually problems were not specified in the results unless selected by at least two or three percent of the population. Proportions lower than this were treated as if they were selected by none of the population. Obviously this method will not denote the periods and intensities of public concern with great accuracy, but it does provide a useful approximation. See Stroman (1978) for a complete catalog of trends.

7. It is easy, of course, to look back over the course of controversy and "fit" its peaks with various larger issues which were prominent at the time, since there is no shortage of candidate issues. To avoid the worst excesses of post hoc theorizing, I have set some methodological rules for the identification of these larger pertinent issues. First, I consider only those issues prominent enough to be registered on the Gallup poll of "most important problem," which specifies only five to ten issues for a given sampling. Second, I require an obvious connection between the national issue and the technical contro-

versy. Thus, an energy crisis is obviously related to nuclear power but not to fluoridation or the ABM. Third, I require political compatibility between the technical controversy and the national issue. Thus, national concern about communist subversion is an issue of the political right and cannot be used to explain the activities of antinuclear leaders on the political left.

8. I could not find good time-series data for an analysis of changing issue content during the ABM controversy. I did examine 31 feature articles in the *Bulletin of the Atomic Scientists* (1964-74) which were primarily concerned with the ABM, coding them simply as to whether or not they mentioned Vietnam, which was not manifestly relevant to the ABM debate. Vietnam is mentioned in none of the three articles published before 1967, in 13 of 25 articles published in 1967-70 (52 percent), and in none of the three articles published in 1971.

# 9
# Controlling Technology

Public opposition to technology is not new. The contentious histories of the lightning rod and of the Luddites who smashed the machines that were to put them out of work, as well as of vaccination, the telegraph, the railroad, the automobile, the use and control of nuclear power after World War II, fluoridation in the 1950's, and others, tell us that technical controversies have been around a long time, if not in the strength and numbers we see today. They have become more common since 1970 and more salient, particularly the giant controversy over nuclear power.

## Why Are There More Protests?

Some analysts believe that these controversies represent a basic loss of confidence in science and scientific reasoning, a loss of faith in the ideal of progress through economic growth and prosperity (Kantrowitz, 1978; Nisbet, 1979). But to argue that those who oppose nuclear power are at the same time opposing progress is to accept completely the proponents' claim that nuclear power is an important requirement for progress. This interpretation ignores the fact that most opponents do not believe the industry's forecasts of economic disaster unless we have a nuclear power program. The antinuclear ideology holds that important social benefits can be maintained, without nuclear power, through conservation, the development of renewable energy sources (particularly solar), and by matching power sources to end uses. This ideology does not envision a degradation of life style but, to the contrary, promotes increases in health, decreases in environmental insult, and a spread of benefits to those portions of the population which do not now enjoy them. This seems closer to a Utopian than an antiprogress position. Whether or not this antinuclear program is realistic, it should be clear that it does not oppose progress and prosperity per se. Similarly, the challengers in most other technical controversies, including those where economic growth is not an issue, usually believe that progress and prosperity will be better served if the technology is not implemented than if it is. For these reasons, I do not believe that technical controversies are caused to any great extent by a disdain for progress.

World War I demonstrated both the wonders and horrors of mechanized warfare, and since then Western attitudes toward science and technology have best been characterized as ambivalent, praising and damning at the same time. While the decades between the world wars showed a great delight with technology, as in the Art Deco movement in architecture and product design, the same period was rich in criticism of science and technology (Hughes, 1975), producing classics like Aldous Huxley's *Brave New World* (1932) and Charlie Chaplin's *Modern Times* (1936). After World War II it was impossible to doubt the power of science-based technology, or to believe that it was necessarily progressive and good. I grew up with the science fiction of that time in which mutant monsters, produced inadvertently by nuclear radiation, were beaten at the end of the movie by a brave young scientist and a beautiful girl who was the daughter or assistant of a kind and wise older scientist. This ambivalence toward science and its products continues into the 1980's, now more salient because there is more technology in our lives and because we are more interested in it and knowledgeable about it. If one focused exclusively on the pessimistic articles and films produced in the last decade, the accidents involving jumbo jets, nuclear power, chemical storage, and fallen satellites, then a negative impression would be inevitable. On the other hand, if one focused on achievements in space, microprocessors, and medical technology, then the impression would be positive. Most members of the public apparently recognize both good and bad features of technology, combining them into a net view that is positive, if qualified (La Porte and Metlay, 1975; Marshall, 1979).

Nisbet (1979) believes that it is the intellectuals, if not the majority of the people, who have lost faith in progress and scientific reasoning, and he is no doubt correct to some extent. However, it is also true that some of the leading opponents of particular technologies are themselves prominent and productive scientists, and many of those who oppose one technology are active proponents of another (Chapter 5). Furthermore, scientific reasoning is used by *both* sides in a controversy, not just those in favor of the technology (Chapter 2). A healthy distrust of scientists appears on both sides too, each doubting the other's experts. Unfortunately, we also see some partisans on each side who reject immediately *anything* that is said by an opposing scientist while accepting uncritically the claims of their own experts. This is not a general skepticism about expertise but simply blatant bias in deciding whom to believe and whom not to. These observations do not seem to me to be consistent with the claim that opposition to a technology basically reflects a loss of faith in science.

I do not believe that the recent increase in technical controversies is properly explained by changing American beliefs about science or progress. A better explanation begins with the recognition that opposition to a technology is a special case of a broader class of political activities usually referred to as "special interest" politics, as opposed to the politics of party identification or patronage (McFarland, 1976). Since the 1960's, there has been a marked increase in the number of groups concerned about particular issues such as peace or arms, schools, pollution, abortion, or electoral reform. Encouraged by legislative actions and administrative rulings, the number of interest-group political-action committees quadrupled between 1972 and 1979 (Broder, 1979). The growth of this form of political activity is not completely understood, but some contributors to growth are apparent. One is growth in the college-educated proportion of the population. (Education is known to be correlated with political participation [Verba and Nie, 1972].) Another is the model provided by popular attempts to improve civil rights, to end the war in Vietnam, to clean up the environment, and to depose a president. Legislative actions, court rulings, and administrative decisions have given new power to challenge groups, for example, the creation of environmental impact statements for federally funded projects, and the opportunities for successful suits based on this requirement. These factors have encouraged a growth in special interest politics which is not limited to the left or right, nor to any particular issue content, and it is certainly not limited to problems of technology.

Organizations involved with these activities grew and became established as effective adversaries, as in the cases of Common Cause, the Sierra Club, and the Nader groups. They attracted talented and enthusiastic personnel, cash resources, and sympathetic (or at least interested) mass media, all enhancing their ability to pursue both technical and nontechnical issues (McFarland, 1976). The growth of antitechnology protests need not be explained as a unique phenomenon but as simply one component of the large and growing body of special interest politics.

## Knowledge and Control

Technical controversies are part of the normal flow of political activity. Therefore it seems reasonable to decide the political issues raised in these controversies in the same way we decide other political issues. *If* we assume that our current political system is an acceptable means for making nonscience policy—a strong assumption—then it ought to be acceptable for making policy regarding science and technology, with little modification. The counterargument to this position is that science and technology are special and therefore must be treated by special means. The particular ways in which they are regarded as special are, first, that they carry their own inertia and so are impervious to social control, and second, that their subject matter is beyond the comprehension of laymen and therefore policy must be made by technical experts. Are either of these points valid?

In *Frankenstein,* Mary Shelley (1831) wrote of a scientist who lost control of his discovery and was eventually destroyed by it, an image that is easily transferred to today's atomic scientists. Technology too is often portrayed as a force that pushes forward inexorably, driven by its own momentum, choosing its own direction, and accelerating unperturbed by any human efforts to control it (Winner, 1977). Perhaps these ideas had more force in the 1950's and 1960's when all "ripe" technologies, whether in warfare, space, or medicine, seemed to be running at full speed. Today things have slowed down, research funds are not so readily available, and we have examples of large technologies that have been stopped in their tracks. We know now that there are two ready levers to control the pace and direction of science and technology: money and government regulation. These are traditional mechanisms which require no special system of political control. We now understand that the usual political bodies—the Executive Branch, Congress, and the courts and state agencies—can increase or decrease the flow of money, and they can tighten or loosen regulatory requirements. We can turn on a war on cancer or turn off space exploration by shifting federal funds, though of course the funding of increased research on, say, cancer does not guarantee that completely satisfactory results will be achieved. We may set standards for the safety of nuclear power plants or the permissible levels of automobile emissions, though there will be instances when these standards are seriously violated. Nonetheless, gross controls exist and are currently in practice.

Our problem is not the absence of powerful controls but rather the failure to exercise properly the controls which are available. This difficulty extends to many policy areas besides science and technology, but in this case the problem is compounded because the material is difficult to understand, the esoteric business of specialists. How should we deal with this problem of specialized knowledge in making technology policy?

It is sometimes claimed that the average citizen is capable of understanding most technical issues that are relevant to policy decisions (e.g., Casper, 1976). I don't believe that anyone who has tried to teach statistics to humanities doctoral students would hold that view for long. The typical voter in a referendum on fluoridation or nuclear power seems to have at best a vague understanding of the underlying scientific disputes which have been discussed during the campaign. The confusion of legislators, when hearing technical experts make apparently inconsistent claims about factual matters, is apparent from congressional hearings (Chapter 2).

One could simply ignore technical issues, realizing that controversies are basically political in nature and are rarely settled by the resolution of factual disputes (Chapter 5). Some courts follow essentially this course today when hearing technical cases. These come up typically when a regulatory agency has set a standard on, say, effluent of a carcinogen from a chemical plant. The agency is often sued, either by the plant, which may think the regulation too strict, or by environmentalists who think it too lax (de Nevers, 1973). In either case, the suit goes to court where conflicting technical claims are heard about "safe" levels of population exposure to the carcinogen. The judges know little of technical matters (Thomas, 1977) and usually base their decision on procedural matters, asking if the regulatory agency followed the proper form and acted within its mandate in setting the standard (Bazelon, 1979). Technical policy is thus set without much regard for the substantive issues in dispute.

A more informed approach is to have scientists, who understand the technical issues, settle the policy questions as well. The National Academy of Sciences makes policy recommendations on many technical problems every year, and though these are not binding in law, they have a good deal of stature and frequently become the legal policy of agencies or the Congress. The problem here is that in the process of giving the scientist a strong voice in his arena of competence, we also give him a particularly strong voice in setting policy for the society, an arena in which he has neither special competence nor a public mandate. Scientists have their own axes to grind, and the elite scientists of the Academy have elite axes.

The approach which I prefer is to separate scientific questions of fact from policy questions of value, a separation which seems quite feasible (Chapter 3). This is a premise of the science court as well as of other proposals for dealing with technical disputes. Whichever of these mechanisms one uses for examining facts, the basic principle is to let the scientists do what they are especially well equipped to do, which is science, while keeping them away from policy decisions, where they have neither special competence nor the public mandate. Once the factual matters are settled, at least provisionally, then the usual procedures of government can be used to make policy decisions, whether by the Congress, the federal agencies, the courts, or by citizens voting in a referendum.

My preference runs counter to the current call for extraordinary public participation in policy decisions regarding science and technology. At one time it was commonly thought that the wisest solutions to technical problems would come from interdisciplinary congresses of scientists, engineers, philosophers, humanists, and clergymen, all pooling their diverse perspectives into a synergistic whole that would somehow be greater than the sum of its parts. In time this view was recognized as elitist (laborers and housewives were never included), and there were new suggestions that the participation of the common man and woman was more important than the involvement of philosophers or clerics. This view has received support from "public interest" groups, though they represent the average citizen about as well as corporate boards represent small stockholders. When President Carter appointed a special commission to investigate the accident at Three Mile Island, he included a "housewife and mother of six" who lived near the site. When the City Council of Cambridge challenged the construction of a laboratory at Harvard for recombinant DNA research, it appointed a much-publicized "citizen court" of eight lay persons from diverse occupations and educational levels (Krimsky, 1978), which presumably was a better representative of the city than the City Council. Sweden, the Netherlands, and Austria have made elaborate attempts to involve the average citizen in decision making about nuclear power (Nelkin and Pollak, 1977). All of these procedures run against my suggestion that *normal* political means be used to make science and technology policy (with some modification to accommodate the specialized knowledge problem).

I certainly do not oppose public participation as such, but I do not see why there should be more of it in science and technology policy than in other kinds of policy. We never make a point of bringing housewives and blue collar laborers into formal decisions about the prime interest rate or on whether or not to attack Iran, so why do it when evaluating nuclear power plants and recombinant DNA laboratories? If decision making by government officials is adequate for military and economic policy, then why not for science and technology policy?

## The Proper Function of Technical Controversies

Technical controversies tend to be regarded as aberrations, as undesirable disruptions which ought to be ignored, disposed of, or avoided altogether. However, there is another viewpoint which emphasizes their positive function.

This viewpoint begins with the assumption that the growth of both scientific knowledge and technological invention is good and ought to be encouraged. At the same time, however, great care must be taken in the distribution of scarce research resources and in the implementation and deployment of innovations. We must minimize waste, health risks, environmental insults, and other costs. These are nearly unassailable assumptions, accepted consensually. It follows that there must be proper roles in the society for people who will advance science and promote new technologies.

It would be nice if the scientist and the engineer were cognizant of, and deeply concerned about, the potential risks, secondary consequences, and other costs of their creations. However, it does not seem realistic to me to expect inventors and promoters of innovation to behave this way. The promoters of thymic irradiation for children thought they were performing a great service, saving many lives; they were not convinced by arguments that "enlarged" thymus was innocuous, nor by early indications that such treatment might be dangerous. Of course they took care to avoid obvious problems like burns from x-ray, but the promoter's role is not compatible with cautious restraint unless there is an obvious reason for it.

If we are to have social roles which promote potentially hazardous innovations, we should have other social roles, filled by different people, to counterbalance the promoters, to search for the hazards and other costs which the promoters do not see, or ignore, or actively avoid.

Regulatory agencies fulfill some of this function, but neither the routine work setting of a bureaucracy nor the job motivations of career bureaucrats are conducive to an enthusiastic and imaginative attack on the problem. Furthermore, innovations, by their nature, often fall outside of the competence and mandate of preexisting agencies.

The congressional Office of Technology Assessment was established to deal with some of these concerns, but it too is a bureaucracy and may not be an appropriate setting for the task. Committees of the National Academy of Sciences have been useful in evaluating some hazards which have already been identified, but if we look at salient cases such as nuclear power or DDT, the Academy has not been particularly effective in locating problems. Furthermore there have been occasional charges that members of Academy committees sometimes have vested interests in the technologies which they are supposed to evaluate (e.g., Schiefelbein, 1979; Boffey, 1975).

There have by now been numerous instances when the informal process of social controversy has been more effective in identifying and explicating the risks and benefits of a technology than have been any of the formal means which are supposed to do this. Critics attack with great vigor, stretching their imaginations for all manner of issues with which to score points against their target. Proponents counterattack, producing new analyses and funding new experiments in order to refute the critics. Each side probes and exposes weaknesses in the other side's arguments. As the controversy proceeds, there is a filtering of issues so that some with little substance become ignored while others move to the fore.

The coming and going of issues is particularly clear in the long controversy over nuclear power. I examined polemic literature against nuclear power from the period 1968-72, comparing it to post-Arab oil embargo literature, and could easily discern the shifting pattern of major concerns from one period to another. Here are problems which were perceived as major in 1968-72 and then became minor afterwards: (1) "thermal pollution" of lakes and rivers by waste heat from the reactor; (2) cancers and genetic damage resulting from routine radiation released by normally operating power plants; (3) the Price-Anderson Act, which provided federally subsidized insurance for the nuclear industry as well as limits on liability; and (4) the poor performance of the Atomic Energy Commission (AEC) as a regulatory agency.

Why do some issues fade away while others persist? Perhaps some problems disappeared because they were solved. Early criticism of the AEC's performance—that it could not simultaneously promote and regulate nuclear power—may have been settled by the dissolution of the AEC and its replacement by separate promotional and regulatory agencies. But virtually all the critics of the AEC, and many of its supporters, regarded the reorganization as a superficial shuffle of AEC personnel. Yet the issue of regulatory competence declined in salience, at least until 1979 when it reemerged as a reaction to the accident at Three Mile Island.

The decline of concern about routine radiation emissions from normally operating power plants is usually explained as a result of the stringent revision by the AEC of its radiation standards in 1971. Declining concern with thermal pollution is similarly explained as the result of the increased insistence of the AEC on cooling towers and ponds at that time. But the early 1970's also marked the decline of massive popular interest in the environment, so the passing of these particularly ecological problems may simply reflect waning public concern with the larger issue of the environment, as I suggested in the last chapter. I see no indication that the insurance issue has been settled; it seems simply to have diminished.

New issues have emerged since 1972. One of the most exciting is the problem of safeguarding nuclear material from terrorists who might construct a bomb. Certainly this issue parallels the fascination of the public during the 1970's with terrorist activities, more than it reflects any new facts about bomb construction. After 1974, when India exploded an atomic bomb built from fissionable material produced in a Canadian-supplied reactor, critics brought the nuclear proliferation issue into the controversy.

The Arab oil embargo of 1973 initiated the energy crisis which has been with us in more or less intensity since then. In the ensuing climate of public opinion, anyone who opposed nuclear energy was likely to champion an alternate energy source (particularly small-scale solar), to emphasize the need for energy conservation, and to minimize our need for additional energy. These issues are prominent in the polemic literature after 1973 though they were virtually non-existent before then.

What of the more persistent issues? There are two such issues, that of catastrophic nuclear reactor accidents, and of disposal of long-term radioactive wastes. These continue as major concerns of the critical literature throughout the controversy. Perhaps the test of time separates wheat from chaff, and these are the "real" issues. Yet even in the persistent concern with reactor accidents, we see a shift in emphasis from year to year: first emphasis was laid on human error, then in 1971-74 on the unproven emergency core-cooling system, then on the credibility of the "Rasmussen Report" (Rasmussen, et al., 1975), which purports to estimate the probability and severity of various kinds of reactor accidents, and most recently, after Three Mile Island, back to human error and regulatory competence.

Precisely the same kind of shifting occurs in promotional rhetoric. The nuclear industry has extolled atomic power for 25 years, but its major advantage has changed repeatedly. In the beginning it was touted as the electricity that would be too cheap to meter, which proved to be an exaggeration. With the rise of environmental concern in the 1960's, it became the clean source of electricity. By 1970, failing to win the support of environmentalists, the industry supported a number of studies showing that nuclear power was superior to fossil fuel generation in risk-benefit tradeoffs. After the oil embargo of 1973, nuclear power was the only proven energy source that would end our dependence on Arab oil.

New issues are continually thought up by partisans, and others are thrust at us by events such as the Arab oil embargo or the accident at Three Mile Island. Some issues quickly fall of their own weight, such as the claim that nuclear power facilities consume more energy than they produce; other issues never gain any following, such as the claim that increasing radiation levels have caused mental deficiencies in children, thus accounting for the decade-long decline in academic achievement test scores.

It would be a mistake to view this process as nothing more than the shifting of popular concern from one faddish issue to another. It is true that some unimportant issues gain an inordinate amount of attention for awhile, but they usually drop from concern once the popular mood changes, unless they can sustain interest on their intrinsic merits. In the meantime, intelligent people on both sides of the controversy search enthusiastically for new problems, diligently preparing charges and rebuttals, testing the strength of their arguments in open debate. In this manner the controversy functions as a funnel, bringing diverse problems together, and it has the potential to act as a sieve which separates important concerns from those without real merit.

One should have no illusions that this process will lead to a settlement of the controversy. We sometimes hear people speak as if a controversy is a set of $n$ issues such that, if each were solved, the controversy would end. That is not the nature of political debate. The alignment is primary and the various rationales for it—the substantive issues—are often secondary, coming and going while the basic alignment persists. But that need not be a problem for the policy maker who understands the dynamics of controversy and who recognizes that the proper function of a controversy is the identification and evaluation of potential problems, as an informal method of technology assessment.

Controversies bring their share of problems. In delaying the implementation of a technology, they may deny to society important benefits, at least for awhile, as in the case of the lightning rod. It is difficult to judge whether we take a net gain or loss from such delays. Our children have more cavities than they would have if the whole nation had been fluoridated in the 1950's, but most of us do not regret that the SST was stopped, and no doubt England and France wish that someone had stopped their's as well. Research on recombinant DNA has probably not lost much ground, in the long view, because of a pause for evaluation. Perhaps we will suffer from the slowdown in nuclear power-plant production, though one might speculate that if the costs of a slowdown become large, the building program will speed up in spite of some objections.

Another problem with technical controversies is that they are chaotic, and therefore we fail to achieve the potential benefits which are available to us. In a controversy each side is ultra motivated to make the strongest case it can while at the same time finding flaws in the other side's position. When these arguments are juxtaposed, as I have attempted to do in Chapter 2, flaws and polemic tricks are disposed of, and we see the beauty of first-class reasoning in support of one's own ends. Unfortunately, in most controversies the adversaries never confront one another as I have brought them together on paper. They address different audiences, or the same audience at different times or in formats which allow them to speak past one another. We become confused when one expert seems to contradict the other, and we cannot locate the reason for their disagreement. The crux of technical disputes between experts may stand out clearly in Chapter 2, but it certainly was not clear to me or my colleagues when I began that analysis. If we could bring some order to the chaos, so that we might better understand the bases for these disagreements, then the value of such arguments would be greatly enhanced.

A final problem with technical controversies is that it is too difficult to start one. I have a number of students deeply concerned about the problems of a technological society, energy issues being at the top of their lists. As they examine the various energy options, they see problems and benefits associated with each one. Yet there is only one energy option which they can do much about, and that is nuclear power, which has readily available channels for opposition right on campus. They have no potent way to express their concerns about the explosive potential of liquified natural gas, about the pollution of coal, or about the hazards of any energy source except nuclear power. It is the only protest in town; you either join it or go home and study.

The power to start a serious and credible technical controversy lies in few hands, notably the environmental and consumer groups, and some prestigious scientists who have good access to the nation's mass media (Mazur, 1981b). Controversies which come from less orthodox sources have trouble gaining credibility in higher circles. The fluoridation controversy came from the grassroots and was never taken seriously in the academic community. I suggest that if most scientists looked today at the evidence that was available on the safety of fluoridation back in 1950, when the drive for national implementation began, they would agree that the decision was premature if not foolhardy. Yet antifluoridationists were lumped together as cranks and kooks, and social scientists called their arguments "antiscience." Nonetheless, many communities rejected fluoridation, putting the proponents on the defensive. They had to strengthen their case for the presence of benefit with minimal risk, and I believe that an impressive body of evidence to that effect now exists.

Every day my family drinks fluoridated water and uses fluoridated toothpaste without worry. About once a year I accept our dentist's judgment that the children's teeth should have a topical application of fluoride. It seems an innocuous procedure, and I suppose my parents felt the same way when a physician suggested that my chronically infected tonsils be irradiated. I do not remember that physician, but I doubt that he was any less cautious than our dentist. Both the working dentist and the working physician treat their patients according to the fashion in their professions at that time. Sometimes the fashions are unwise, and if so, it is best that we learn of it sooner rather than later, even at the expense of some controversy.

# References

Adams, B. 1971. *Ballistic Missile Defense*. New York: Elsevier.

Alexanderson, E., and Wagner, H., eds., 1979. *Fermi-1: New Age for Nuclear Power*. LaGrange Park, Ill.: American Nuclear Society.

Alpern, D.; Bishop, J.; and Cook, W. 1975. Pulling the Plug on A-Power. *Newsweek* (Feb. 24): 23.

Alpern, D.; Reese, M.; and Walcott, J. 1978. Anti-atom Alliance. *Newsweek* (June 5): 27, 29.

American Dental Association. 1965. Comments of the Opponents of Fluoridation. *Journal of the American Dental Association* 71: 1156.

Ames, M. 1978. *Outcome Uncertain*. Washington, D.C.: Communications Press.

Armstrong, W.; Bittner, J.; and Treloar, A. 1954. Untitled testimony. In *Hearings on H.R. 2341 before the Committee on Interstate and Foreign Commerce: Fluoridation of Water*, pp. 307-9. 83 Congress, session 2. Washington, D.C.

Barkan, D. 1979. Strategic, Tactical, and Organizational Dilemmas of the Protest Movement Against Nuclear Power. *Social Problems* 27 (Oct.): 19-37.

Barnes, J. 1929. Late Effects of Treatment of Thymus. *American Journal of Roentgenology* 22: 220-25.

Bazelon, D. 1979. Risk and Responsibility. *Science* 205: 277-80.

Berger, J. 1977. *Nuclear Power: The Unviable Option*. Revised edition. New York: Dell.

Boffey, P. 1970. Gofman and Tamplin: Harassment Charges Against AEC, Livermore. *Science* 169: 838.

———. 1975. *The Brain Bank of America*. New York: McGraw-Hill.

———. 1976. Nuclear Power Debate: Signing Up the Pros and Cons. *Science* 192: 120-22.

Bond, V. 1970. *Radiation Standards, Particularly as Related to Nuclear Power Plants*. Raleigh, N.C.: Council for the Advancement of Science Writing.

Boyd, E. 1927. Growth of the Thymus. *American Journal of Diseases of Children* 33: 867-79.

134     *References*

Brehm, W., 1924. Sclerotizing Strumitis with Compression of the Larynx as a Late Roentgen Injury Following Thyroid Irradiation. *American Journal of Roentgenology* 12: 586.

Broder, D. 1979. PAC Power Curbed. *Syracuse Post-Standard* (Oct. 27): 4.

Brooks, R. 1978. A Minnesota Background. Meeting of the American Bar Association, August 4. New York.

Brush, S. 1974. The Prayer Test. *American Scientist* 62: 561-63.

Cahn, A. 1974. American Scientists and the ABM. In *Scientists and Public Affairs*, ed. A. Teich, pp. 41-120. Cambridge: MIT Press.

Carman, R., and Miller, A. 1924. Occupational Hazards of the Radiologist. *Radiology* 3: 408-19.

Carson, R. 1962. *Silent Spring*. Boston: Houghton Mifflin.

Carter, L. 1979. The 'Movement' Moves on to Antinuclear Protest. *Science* 204: 715.

Casper, B. 1976. Technology Policy and Democracy. *Science* 194: 29-35.

Clark, D., 1955. Association of Irradiation with Cancer of the Thyroid in Childhood and Adolescence. *Journal of the American Medical Association* 159: 1007-9.

Cohen, I., 1952. Prejudice Against Lightning Rods. *Journal of the Franklin Institute* 253: 393-440.

Cole, J., and Cole, S. 1973. *Social Stratification in Science*. Chicago: University of Chicago Press.

Coleman, J. 1957. *Community Conflict*. New York: Free Press.

Commoner, B. 1979. *The Politics of Energy*. New York: Alfred Knopf.

Conti, E., and Patton, G. 1948. Study of the Thymus in 7,400 Consecutive Newborn Infants. *American Journal of Obstetrics and Gynecology* 56: 884-92.

Coser, L. 1956. *The Functions of Social Conflict*. New York: Free Press of Glencoe.

Crain, R.; Katz, E.; and Rosenthal, D. 1969. *The Politics of Community Conflict*. Indianapolis: Bobbs-Merrill.

Curtis, R., and Hogan, E. 1969. *Perils of the Peaceful Atom*. New York: Ballantine.

Davies, J. 1962. Toward a Theory of Revolution. *American Sociological Review* 27: 5-19.

———. 1969. The J-curve of Rising and Declining Satisfactions as a Cause of Some Great Revolutions and a Contained Rebellion. In *Violence in America*, eds. H. Graham and T. Gurr, pp. 671-709. New York: Bantam Books.

Davis, J.; Smith, T.; and Stephenson, C. 1980. *General Social Surveys, 1972-80: Cumulative Codebook*. Chicago: National Opinion Research Center.

Davis, M. 1956. Community Attitudes toward Fluoridation. *Public Opinion Quarterly* 23: 474.

Dean, H. 1938. Endemic Fluorosis and its Relation to Dental Caries. *Public Health Reports* 53: 1443-52.

DeGroot, L., and Paloyan, E. 1973. Thyroid Carcinoma and Radiation. *Journal of the American Medical Association* 225: 487-91.

de Nevers, N. 1973. Enforcing the Clean Air Act of 1970. *Scientific American* 228 (June): 14-21.

Dibner, B. 1977. Benjamin Franklin. In *Lightning,* vol. 1, ed. R. Golde. New York: Academic Press.

Doff, L. 1970. *Atomic Power and the Public Mind.* New York: Atomic Industrial Forum.

Donaldson, S. 1930. Hyperplasia of the Thymus. *American Journal of Roentgenology* 24: 523-33.

Douvan, E., and Withey, S. 1954. Public Reaction to Non-military Aspects of Atomic Energy. *Science* 119: 1-3.

Duffy, B., and Fitzgerald, P. 1950. Thyroid Cancer in Childhood and Adolescence; A Report on 28 Cases. *Cancer* 3: 1018-32.

Duncan, O.D. 1978. Sociologists Should Reconsider Nuclear Energy. *Social Forces* 57: 1-22.

Ebbin, S., and Kasper, R. 1974. *Citizen Groups and the Nuclear Power Controversy.* Cambridge: MIT Press.

Electric Light and Power Companies, 1971. Untitled advertisement, *Time* (Dec. 20).

*Electric Power and the Atom,* undated. Minneapolis: Northern States Power Co.

Evans, R. 1966. The Effect of Skeletally Deposited Alpha Emitters in Man. *British Journal of Radiology* 39: 881-95.

Exner, F.; Waldbott, G.; and Rorty, J. 1957. *The American Fluoridation Experiment.* New York: Devin-Adair.

Farhar, B.; Weis, P.; Unseld, C.; and Burns, B. 1977. *Public Opinion About Energy: A Literature Review.* Golden, Co.: Solar Energy Research Institute.

Faulkner, P. ed., 1977. *The Silent Bomb.* San Francisco: Friends of the Earth.

*Fluoridation Census 1969.* 1970. Bethesda, Md.: U.S. Dept. of Health, Education, and Welfare.

Frankel, H. 1981. The Paleobiogeographical Debate Over the Problem of Disjunctively Distributed Life Forms. *Studies in History and Philosophy of Science,* in press.

Friedlander, A. 1907. Status Lymphaticus and Enlargement of the Thymus: with a Report of a Case Successfully Treated by the X-ray. *Archives of Pediatrics* 24: 490-501.

———. 1911. Involution of the Thymus by the X-ray. *Archives of Pediatrics* 28: 810-29.

Fuller, J. 1975. *We Almost Lost Detroit.* New York: Ballantine.

Gamson, W. 1961a. Public Information in a Fluoridation Referendum. *Health Education Journal* 19: 47.

———. 1961b. The Fluoridation Dialogue: Is it an Ideological Conflict? *Public Opinion Quarterly* 25: 526.

————. 1966. Reputation and Resources in Community Politics. *American Journal of Sociology* 63: 121.

————. 1968. *Power and Discontent.* Homewood, Ill.: Dorsey Press.

————. 1975. *The Strategy of Protest.* Homewood, Ill.: Dorsey Press.

Gerlach, L. 1978. The Great Energy Standoff. *National History* 87: 22-32.

Gerlach, L., and Hine, V. 1970. *People, Power, Change: Movements of Social Transformation.* Indianapolis: Bobbs-Merrill.

Gillette, R. 1971. Nuclear Reactor Safety: A Skeleton at the Feast? *Science* 172: 918-19.

Gofman, J., et al., 1971. Radiation As an Environmental Hazard. Mimeo. Livermore, California: Lawrence Radiation Laboratory.

Gofman, J., and Tamplin, A. 1971. *Poisoned Power.* Emmaus, Pa.: Rodale Press.

Goldstein, G., and Mackay, I. 1969. *The Human Thymus.* London: William Heinermann Medical books.

Goodell, R. 1977. *The Visible Scientists.* Boston: Little, Brown & Co.

Gray, C. 1975. Hawks and Doves: Values and Policy. *Journal of Political and Military Sociology* 3: 85-94.

Green, A. 1961. The Ideology of Anti-fluoridation Leaders. *Journal of Social Issues* 17: 13.

Greenberg, D. 1967. *The Politics of Pure Science.* New York: New American Library.

Grobstein, C. 1977. The Recombinant - DNA Debate. *Scientific American* 237 (July): 22-33.

Groth, E., III. 1972. Untitled article. *National Fluoridation News* (Jan.-March), Hempstead, N.Y.: 2-4.

————. 1973. *Two Issues of Science and Public Policy.* Ann Arbor: University Microfilms.

————. undated. *Results of Opinion Survey on Attitudes of Opponents of Fluoridation.* Pasadena, Cal.: Caltech Population Program.

Gurr, T. 1970. *Why Men Rebel.* Princeton: Princeton University Press.

Gwynne, P.; Bishop, Jr., J.; and Michaud, S. 1976. How Safe is Nuclear Power? *Newsweek* (April 12): 70-75.

Gyorgy, A., and friends, 1979. *No Nukes.* Boston: South End Press.

Halsted, T. 1971. Lobbying Against ABM: 1967-1970. *Bulletin of the Atomic Scientists* 27: 23-28.

Harris, L., and Associates, 1976. *A Second Survey of Public and Leadership Attitudes toward Nuclear Power Development in the United States.* New York: Ebasco Services.

Hayes, D. 1978. The Cambridge Story. Meeting of the American Bar Association, August 4. New York.

*Hearings Before the House Select Committee to Investigate the Use of Chemicals in Foods and Cosmetics,* 82 Congress, session 2, part 3, 1952. Washington, D.C.: U.S. Government Printing Office.

Hempelmann, L. 1968. Radiation-induced Thyroid Neoplasms in Man. In *Thyroid Neoplasia*, eds. S. Young and D. Inman, pp. 267-76. London: Academic Press.

Hensler, D., and Hensler, C. 1979. *Evaluating Nuclear Power: Voter Choice on the California Nuclear Energy Initiative.* Santa Monica: Rand.

Hess, G. 1927. Possible Developmental Defects Following Over-radiation of Thymus in Early Infancy. *Radiology* 9: 506-9.

Hibbs, D. 1973. *Mass Political Violence.* New York: Wiley.

Hoffer, E. 1951. *The True Believer.* New York: Harper.

Holcomb, R. 1970. Radiation Risk: A Scientific Problem? *Science* 167: 854.

Holden, C. 1979. Barry Commoner as First Citizen? *Science* 205: 172.

Horwitch, M. 1978. The American SST. In *Macro-Engineering and the Infrastructure of Tomorrow*, eds. F. Davidson, L. Giacoletto, and R. Salkeld, pp. 139-176. Boulder, Co.: Westview Press.

Hudson, H. 1935. The Thymus Superstition. *New England Journal of Medicine* 212: 910-13.

Hughes, T., ed., 1975. *Changing Attitudes toward American Technology.* New York: Harper and Row.

*INFO*, various dates. Washington, D.C.: Atomic Industrial Forum.

Jenkins, J., and Perrow, C. 1977. Insurgency of the Powerless: Farm Worker Movements (1946-1972). *American Sociological Review* 42: 249-68.

Jerard, E. 1968. *The Case of the Protected Pollutant.* New York: Independent Phi Beta Kappa Environmental Study Group.

Johnston, H. 1976. The Ozone Controversy. *Science* 191: 1125-26.

Jopling, D. 1970. *The Politics of Nuclear Reactor Siting.* Unpublished MBA thesis. Austin: University of Texas.

Kantrowitz, A. 1978. A Technologist Looks at Anti-Technology. Messenger Lectures. Cornell University.

Kash, D.; White, I.; Reuss, J.; and Leo, J. 1972. University Affiliation and Recognition: National Academy of Sciences. *Science* 175: 1076-87.

Kasperson, R.; Berk, G.; Pijawka, D.; Sharaf, A.; and Wood, J. 1980. Public Opposition to Nuclear Power on the Eve of the 1980's. *Science, Technology and Human Values* 5.

Kopp, C. 1979. The Origins of the American Scientific Debate Over Fallout Hazards. *Social Studies of Science* 9: 403-22.

Kornhauser, A. 1959. *The Politics of Mass Society.* Glencoe, Ill.: Free Press.

Kriesberg, L. 1973. *The Sociology of Social Conflicts.* Esglewood Cliffs, N.J.: Prentice-Hall.

Krimsky, S. 1978. A Citizen Court in the Recombinant DNA Debate. *Bulletin of the Atomic Scientists* (Oct.): 37-43.

―――. 1979. Regulating Recombinant DNA Research. In *Controversy: Politics of Technical Decisions*, ed. D. Nelkin, pp. 227-54. Beverly Hills: Sage.

Ladd, E. 1969. Professors and Political Petitions. *Science* 163: 1425-30.

―――. 1970. American University Teachers and Opposition to the Vietnam War. *Minerva* 8: 542-56.

Ladd, E., and Lipset, S.M. 1972. Politics of Academic Natural Scientists and Engineers. *Science* 176: 1091-2011.

———. 1976. *The Divided Academy*. New York: Norton.

Lambright, W.H. 1967. *Shooting Down the Nuclear Airplane*. Indianapolis: Bobbs-Merrill.

———. 1976. *Governing Science and Technology*. New York: Oxford University Press.

LaPorte, T., and Metlay, D. 1975. Technology Observed: Attitudes of a Wary Public. *Science* 188: 121-27.

Lapp, R. 1975. Assessing Nader's Critical Mass. Atomic Industrial Forum, Feb. 5. New York.

Lawless, E. 1977. *Technology and Social Shock*. New Brunswick: Rutgers University Press.

Leahy, P., and Mazur, A. 1978. A Comparison of Movements Opposed to Nuclear Power, Fluoridation, and Abortion. In *Research in Social Movements, Conflict and Change,* vol. 1, ed. L. Kriesberg, pp. 143-54. New York: JAI Press.

———. 1980. The Rise and Fall of Public Opposition to Specific Social Movements. *Social Studies of Science* 10: 259-84.

Lewis, R. 1972. *The Nuclear-Power Rebellion*. New York: Viking Press.

List, D., and Katz, N. 1979. Snapshot from Seabrook: A Profile of Clams in Action, 1978. Association for Voluntary Action Scholars, San Antonio.

Lovins, A., et al., 1977. *Soft Energy Paths: Toward a Durable Peace*. San Francisco: Friends of the Earth.

McCarthy, J., and Zald, M. 1973. *The Trend of Social Movements in America: Professionalization and Resource Mobilization*. Morristown, N.J.: General Learning Press.

McCaughey, R. 1976. American University Teachers and Opposition to the Vietnam War: A Reconsideration. *Minerva* 14: 307-29.

McClure, F. 1970. *Water Fluoridation: The Search and the Victory*. Bethesda, Md.: U.S. Dept. of Health, Education and Welfare.

McCracken, S. 1979. The Harrisburg Syndrome. *Commentary* (June): 27-39.

McFarland, A. 1976. *Public Interest Lobbies*. Washington, D.C.: American Enterprise Institute.

McGuire, P. undated. The Effect of the 765 Line on the Residents and the Quality of Life Near the Line. Mimeo. State University of New York at Morrisville.

McNeil, D. 1957. *The Fight for Fluoridation*. New York: Oxford University Press.

Markle, G., and Petersen, J. 1979. The Laetrile Controversy. In *Controversy: Politics of Technical Decisions,* ed. D. Nelkin, pp. 159-79. Beverly Hills: Sage.

Markle, G.; Petersen, J.; and Wagenfeld, M. 1978. Notes from the Cancer Underground: Participation in the Laetrile Movement. *Social Science and Medicine* 12: 31-37.

Marshall, E. 1979. Public Attitudes to Technological Progress. *Science* 205: 281-85.

Marx, G., and Wood, J. 1975. Strands of Theory and Research in Collective Behavior. *Annual Review of Sociology* 1: 363-428.

Marx, J. 1979. Low-level Radiation: Just How Bad is It? *Science* 204: 160-64.

Mausner, B., and Mausner, J. 1955. A Study of the Anti-scientific Attitude. *Scientific American* 192: 35-39.

Mauss, A. 1975. Social Problems as Social Movements. Philadelphia: Lippincott.

Maynard, W.; Nealey, S.; Hebert, J.; and Lindell, M. 1976. *Public Values Associated with Nuclear Waste Disposal*. Seattle: Battelle.

Mazur, A. 1968. A Nonrational Theory of Conflict and Coalition Formation. *Journal of Conflict Resolution* 12: 196-205.

———. 1973. Disputes Between Experts. *Minerva* 11: 243-62.

———. 1975. Opposition to Technical Innovations. *Minerva* 13: 58-81.

———. 1977a. Public Confidence in Science. *Social Studies of Science* 7: 123-25.

———. 1977b. Science Courts. *Minerva* 15: 1-14.

———. 1981a. TMI and the Scientific Community. *Proceedings of the New York Academy of Sciences,* in press.

———. 1981b. Media Coverage and Public Opinion on Scientific Controversies. *Journal of Communication,* in press.

Mazur, A., and Conant, B. 1978. Opposition to a Local Nuclear Waste Repository. *Social Studies of Science* 8: 235-43.

Melber, B.; Nealey, S.; Hammersla, J.; and Rankin, W. 1977. *Nuclear Power and the Public: Analysis of Collected Survey Research*. Seattle: Battelle.

Metzger, P. 1972. *The Atomic Establishment*. New York: Simon and Schuster.

Metzner, C. 1957. Referenda for Fluoridation. *Health Education Journal* 15: 168.

Milch, J. 1979. The Toronto Airport Controversy. In *Controversy: Politics of Technical Decisions,* ed. D. Nelkin, pp. 49-68. Beverly Hills: Sage.

Miller, R. 1969. Delayed Radiation Effects in Atomic-bomb Survivors. *Science* 166: 572.

Mitchell, R. 1979. The Public Response to Three Mile Island. Discussion paper D-58. Washington, D.C.: Resources for the Future.

———. 1980. Public Opinion and Nuclear Power Before and After Three Mile Island. *Resources* (January-April): 5-7.

Morse, J. 1928. The Thymus Obsession. *Boston Medical and Surgical Journal* 33: 1547-52.

Mueller, J. 1966. The Politics of Fluoridation in Seven California Cities. *Western Political Quarterly* 19: 54.

———. 1968. Fluoridation Attitude Change. *American Journal of Public Health* 58: 1876.

Murphy, J., Witherbee, W.; Craig, S.; Hussey, R.; and Strum, E. 1921. Induced Atrophy of Hypertrophied Tonsils by Roentgen Ray. *Journal of the American Medical Association* 76: 228.

Nader, R. 1965. *Unsafe at Any Speed*. New York: Grossman.

Nelkin, D. 1971. *Nuclear Power and Its Critics*. Ithaca: Cornell University Press.

———. 1972. *The University and Military Research*. Ithaca: Cornell University Press.

———. 1974. *Jetport*. New Brunswick, N.J.: Transaction Books.

Nelkin, D., and Pollak, M. 1977. The Politics of Participation and the Nuclear Debate in Sweden, the Netherlands, and Austria. *Public Policy* 25: 333-57.

———. 1980. *The Atom Beseiged: Extra-parliamentary Dissent in France and Germany*. Cambridge: MIT Press.

Nichols, D. 1974. The Associational Interest Groups of American Science. In *Scientists and Public Affairs*, ed. A. Teich, pp. 123-70. Cambridge: MIT Press.

Nisbet, R. 1979. The Rape of Progress. *Public Opinion* 2: 2-6, 55.

Nolan, C. 1973. Seat Belt Vote Called a Charade. *Syracuse Post-Standard* (April 16): 6.

Novick, S. 1969. *The Careless Atom*. Boston: Dell.

Obershall, A. 1973. *Social Conflict and Social Movements*. Englewood Cliffs, N.J.: Prentice-Hall.

Perrow, C. 1979. TMI: A Normal Accident. In *Social Science Aspects of the Accident at Three Mile Island*, eds. D. Sills, C. Wolf, and V. Shelanski. New York: Social Science Research Council.

Peterson, R., and Miller, N. 1924. Thymus of Newborn and its Significance to the Obstetrician. *Journal of the American Medical Association* 83: 234-38.

Pfahler, G. 1924. The Diagnosis of Enlarged Thymus by the X-ray, and Treatment by X-ray or Radium. *Archives of Pediatrics* 41: 39-46.

Pinard, M. 1963. Structural Attachments and Political Support in Urban Politics. *American Journal of Sociology* 68: 513.

Primack, J., and Von Hippel, F. 1974. *Advise and Dissent*. New York: American Library.

Quimby, E., and Werner, S. 1949. Late Radiation Effects in Roentgen Therapy for Hyperthyroidism. *Journal of the American Medical Association* 140: 1046-47.

Rasmussen, N., et al., 1975. *Reactor Safety Study*. Washington, D.C.: Nuclear Regulatory Commission, document number WASH-1400.

Reed, J., and Wilkes, J. 1980. Sex and Attitudes toward Nuclear Power. American Sociological Association Meeting, August, 1980.

Refetoff, S.; Harrison, J.; Koranfilshi, B.; Kaplan, E.; DeGroot, L.; and Bekerman, C. 1975. Continuing Occurrence of Thyroid Carcinoma after Irradiation to the Neck in Infancy and Childhood. *New England Journal of Medicine* 292: 171-75.

Remsberg, C., and Remsberg, B. 1976. The Hospital that Cared. *Good House-keeping* (May): 101-55.

Reppy, J. 1979. The Automobile Airbag. In *Controversy: Politics of Technical Decisions,* ed. D. Nelkin, pp. 145-58. Beverly Hills: Sage.

Robbins, D., and Johnston, R. 1976. The Role of Cognitive and Occupational Differentiation in Scientific Controversies. *Social Studies of Science* 6: 349-68.

Rosa, E. 1978. The Public and the Energy Problem. *Bulletin of the Atomic Scientists* 34: 5-7.

Royce, P.; Mackay, B.; and DiSabella, P. 1979. Value of Postirradiation Screening for Thyroid Nodules. *Journal of the American Medical Association* 242: 2675-78.

Rubin, D. 1971. *Reporting the Corporate State.* Palo Alto, Cal.: Stanford University thesis.

Sailor, V. 1971. Untitled. *Nuclear Industry* 22: 25.

Sandm'an, P., and Paden, M. 1979. At Three Mile Island. *Columbia Journalism Review* (July/August): 43-58.

Sapolsky, H. 1968. Science, Voters and the Fluoridation Controversy. *Science* 162: 427.

Saunders, I. 1961. The Stages of a Community Controversy. *Journal of Social Issues* 17: 55.

Schiefelbein, S. 1979. The Invisible Threat. *Saturday Review* (Sept. 15): 16-20.

Seaborg, G., and Corliss, W. 1971. *Man and Atom.* New York: Dutton.

Seeley, T.; Fulford, P.; and Treffinger, D. 1971. Research in Public Attitudes toward Nuclear Power and Electric Utilities. Atomic Nuclear Society Meeting, Boston.

Simmel, A. 1961. A Signpost for Research on Fluoridation Conflicts. *Journal of Social Issues* 17: 26.

Simmel, A., and Ast, D. 1962. Some Correlates of Opinion on Fluoridation. *American Journal of Public Health* 57: 1269.

Simpson, C., et al., 1955. Neoplasia in Children Treated with X-rays in Infancy for Thymic Enlargement. *Radiology* 64: 840-5.

Slater, H. 1970. Untitled. *INFO* 32: 2.

Smelser, N. 1962. *The Theory of Collective Behavior.* New York: Free Press.

Smith, A. 1965. *A Peril and A Hope.* Chicago: University of Chicago Press.

Snyder, D., and Tilly, C. 1972. Hardship and Collective Violence in France, 1830 to 1960. *American Sociological Review* 37: 520-32.

Stallen, P., and Meertens, R. undated. *Value Orientations, Evaluations and Beliefs Concerning Nuclear Energy.* Nijmegen, The Netherlands: University of Nijmegen, Dept. of Social Psychology, internal report 77 SO 02.

Starr, C. 1971. The Electric Power Crisis in America. *Look* (Aug. 10): 40.

Sternglass, E. 1972. *Low-level Radiation.* New York: Ballantine.

Steward, M. 1952. Status Lymphaticus. *Proceedings of the Pathological Society* 28:132.

Stroman, C. 1978. *Race, Public Opinion, and Print Media Coverage.* Syracuse: Syracuse University thesis.

Tamplin, A., and Gofman, J. 1970. *Population Control through Nuclear Pollution.* Chicago: Nelson-Hall.

Task Force of the Presidential Advisory Group on Anticipated Advances in Science and Technology. 1976. The Science Court Experiment: An Interim Report. *Science* 193: 653-56.

Taylor, L. undated. What we *Do* Know About Low-level Radiation. *INFO:* 13.

Teich, A., ed., 1974. *Scientists and Public Affairs.* Cambridge: MIT Press.

Thomas, W. 1977. Judicial Treatment of Scientific Uncertainty in the *Reserve Mining* Case. *Proceedings of the Fourth Symposium on Statistics and the Environment.* Washington, D.C.: American Statistical Association; 1-13.

Tilly, C.; Tilly, L.; and Tilly, R. 1975. *The Rebellion Century.* Cambridge: Harvard University Press.

Tripp, A. 1978. A Rebuttal. Meeting of the American Bar Association, August 4. New York.

Tukey, J. 1977. *Exploratory Data Analysis.* Reading, Mass.: Addison Wesley.

Uhlmann, E. 1956. Cancer of the Thyroid and Irradiation. *Journal of the American Medical Association* 161: 504-7.

Ulrich, H. 1946. The Incidence of Leukemia in Radiologists. *New England Journal of Medicine* 234: 45-46.

Union of Concerned Scientists. 1975. *The Nuclear Fuel Cycle.* Cambridge: MIT Press.

Van Liere, K.; Ladd, A.; and Hood, T. 1979. Anti-nuclear Demonstrators: A Study of Participants in the May 6 Anti-nuclear Demonstration. Paper presented 30 October 1979, at the Mid-South Sociological Association, Memphis, Tenn.

Verba, S., and Nie, N. 1972. *Participation in America.* New York: Harper and Row.

Weinberg, A. 1972. Science and Trans-science. *Minerva* 10: 209-22.

———. 1979. Salvaging the Atomic Age. *The Wilson Quarterly* (summer): 88-112.

Whitney, V. 1950. Resistance to Innovation: The Case of Atomic Power. *American Journal of Sociology* 56: 247-54.

Wilson, J. 1973. *Introduction to Social Movements.* New York: Basic Books.

Winner, L. 1977. *Autonomous Technology: Technics-out-of-control as a Theme in Political Thought.* Cambridge, Mass.: MIT Press.

Young, M., and Turnbull, H. 1931. An Analysis of the Data Collected by the Status Lymphaticus Investigating Committee. *Journal of Pathology and Bacteriology* 34: 213-58.

Zald, M., and Ash, R. 1966. Social Movement Organization: Growth, Decay and Change. *Social Forces* 44: 327-40.

Zuckerman, H. 1977. *Scientific Elite.* New York: Free Press.

# Index

Tripp, A., 89, 93
Tukey, J., 108
Turnbull, H., 3

U

Ulrich, H., 6
Union of Concerned Scientists, 70, 71, 74
United Auto Workers, 47
*Unsafe at Any Speed,* 88

V

Van Liere, K., 59
Verba, S., 46, 122
Vietnam, 46, 47, 57, 59-61, 69-80, 113, 115, 119, 122
Von Hippel, F., 70, 71, 87

W

Wagner, H., 104
Wald, G., 73
Warnings, 87-90

Washington-level protests, 86, 90, 91, 94, 95, 101
Watson, J., 62
Weinberg, A., 12, 110
Weisner, J., 87
Werner, S., 6
Wilkes, J., 51
Winner, L., 123
Witherbee, W., 5
Wood, J., 58

X

X-ray therapy, 2-7, 62, 85, 97, 126, 131

Y

Young, M., 3

Z

Zald, M., 99
Zuckerman, H., 80

# About the Author

Allan Mazur is both a sociologist and a technologist. He earned an M.S. in Engineering from UCLA and worked for several years as an aerospace engineer before obtaining a Ph.D. in Sociology from Johns Hopkins University. He has been a member of the social science faculties of MIT and Stanford University, and is currently a professor in Syracuse University's Maxwell School of Citizenship and Public Affairs. A well-known authority on the sociology of technology, he is a frequent lecturer and consultant to government agencies and private groups. He has published widely, including articles in *Science, American Sociological Review, American Journal of Sociology, Social Forces, Minerva,* and numerous other scholarly journals.

# Science/Technology and Public Affairs books from Communications Press, Inc.

*The Dynamics of Technical Controversy.* Allan Mazur
(0-89461-033-3, hardcover; 0-89461-034-1, text pap.)

*Outcome Uncertain: Science and the Political Process.*
Mary E. Ames (0-89461-028-7, hardcover; 0-89461-029-5,
text pap.)

*The Cable/Broadband Communications Book, Vol. 1,
1977-1978.* Edited by Mary Louise Hollowell
(0-89461-027-9, pap.)

*The Cable/Broadband Communications Book, Vol. 2,
1980-1981.* Edited by Mary Louise Hollowell
(0-89461-031-7, pap.)